NETWORKED ENVIRONME
STAKEHOLDER PARTICIPATION IN WATER
RESOURCES AND FLOOD MANAGEMENT

NETWORKED ENVIRONMENTS FOR STAKEHOLDER PARTICIPATION IN WATER RESOURCES AND FLOOD MANAGEMENT

DISSERTATION

Submitted in fulfilment of the requirements of
the Board for Doctorates of Delft University of Technology
and of the Academic Board of the UNESCO-IHE
Institute for Water Education
for the Degree of DOCTOR
to be defended in public on
Tuesday, 18th March 2014, at 10:00 hours
in Delft, the Netherlands

by

Adrian Delos Santos ALMORADIE

Master of Science in Hydroinformatics
UNESCO-IHE Institute for Water Education, Delft, the Netherlands

born in Masbate City, Philippines

This dissertation has been approved by the supervisor:
Prof. dr. D.P. Solomatine

Co-supervisor: Dr. A. Jonoski

Composition of Doctoral Committee:

Chairman	Rector Magnificus Delft University of Technology
Vice-Chairman	Rector UNESCO-IHE
Prof. dr. D.P. Solomatine	UNESCO-IHE / Delft University of Technology, supervisor
Dr. A. Jonoski	UNESCO-IHE, co-supervisor
Prof. D.P. Loucks	Cornell University, New York, USA
Prof. dr. M. Evers	Bonn University, Germany
Prof. dr.ir. N.C. van de Giesen	Delft University of Technology
Prof. dr.ir. A.E. Mynett	UNESCO-IHE / Delft University of Technology
Prof. dr.ir. P. van der Zaag	UNESCO-IHE / Delft University of Technology, reserve member

CRC Press/Balkema is an imprint of the Taylor & Francis Group, an informa business

.
Published by:
CRC Press/Balkema
PO Box 11320, 2301 EH Leiden, The Netherlands
e-mail: Pub.NL@taylorandfrancis.com
www.crcpress.com - www.taylorandfrancis.com

ISBN 978-1-138-02637-7 (Taylor & Francis Group)

To Mylene, Ian Jeremy and Iliana Jianese

To my father Jesus, mother Remedios, siblings Anthony, Aireen and Azenith

To my grandmother Eduvijis and the memory of my grandfather Demosthenes

SUMMARY

Lack of stakeholder awareness, involvement and participation in water resources and flood management (WRFM) planning and decision making processes often creates problems related to the implementation and acceptance of the proposed measures. Stakeholder awareness and participation in disaster prevention and management are crucial and should cover all phases of any disaster event. Moreover, since stakeholders often have a better understanding of the real potential and limitations of their local environment, their involvement in planning and management are of crucial importance.

Stakeholders can be classified in categories, such as government institutes, flood prone communities, NGOs, basin communities, private sector and scientific communities. Information sharing and repeated interaction between stakeholders are needed so as to build trust, negotiate for best possible benefits, and to enhance cooperation across jurisdictions and sectors. The challenge in stakeholders' participation is launching and maintaining the participatory process. Spatial distribution and diverse (even opposed) stakeholders' interests may come as one of the hindrances in maintaining the participatory process.

This research entitled "Networked Environments for Stakeholder Participation (NESP) in Water resources and Flood Management" addresses some of these challenges and hindrances in stakeholder participation. Networked Environments (NE) are web-based computer-aided or mobile environments for remote virtual interaction between participating entities such as stakeholders. NESP is envisioned to enable stakeholder participation in water resources and flood management by providing sharing of information, planning, negotiating and decision support.

The recent advancements of ICT (Information and Communications Technology) provide innovative solutions for the development of the NESP. Since the beginning of the computer and the Internet era, the World Wide Web has been increasingly used in our societies as a technology to access sources of information and for communication among organizations and individuals. Moreover, mobile technology has demonstrated even more advantages for reaching and engaging most of the citizens and potential stakeholders. The advancement of mobile technology and its application development paves a way for its usage for data gathering, remote execution of models and information dissemination. In effect, the Internet, World Wide Web, mobile and wireless technologies present a powerful environment for development and deployment of NESP as envisaged in this study.

The main objective of this work is research into conceptualisation, design and implementation of innovative web-based and mobile environments for stakeholder participation using the recent advanced ICT technologies. It incorporates novel approaches in stakeholder involvement in all phases of project planning and analysis, including negotiation support for deriving options with joint benefits.

Three case specific NESP frameworks were conceptualised to address the three different types of participation: (1) Information and knowledge sharing, (2) Consultative participation and (3) Collaborative decision making. These frameworks were named accordingly based upon the type of participation: (1) NESP-IKS, (2) NESP-CP and (3) NESP-CDM.

The framework termed *NESP-IKS* (**I**nformation and **K**nowledge **S**haring) was conceptualised for an effective assimilation of stakeholders' information and knowledge in WRFM. This can lead to mobilization and utilization of more reliable and up to date information in WRFM processes. Moreover, the framework offers professionals the possibility of using stakeholders' observations to improve their models and forecasts. The conceptual framework has three main components and one optional component: (1) Background information, (2) Information access, (3) Stakeholder participation and (4) Improvement of models and forecasts (*optional*).

The developed framework *NESP-CP* (**C**onsultative **P**articipation) was conceptualised for an effective and more inclusive type of participation. More inclusive participation through consultation can influence existing practices in the management and planning of water resources or floods. The developed NESP-CP application is expected to be a valuable system for awareness raising and stakeholder empowerment in WRFM. The framework has three main components: (1) Risk awareness, (2) Information access and (3) Stakeholder participation.

Lastly, the framework *NESP-CDM* (**C**ollaborative **D**ecision **M**aking) is intended for a collaborative type of participation where stakeholders together with experts identify relevant scenarios and realistic management alternatives that address commonly agreed management objectives. The participating parties subsequently assess the preferred alternatives, first by enabling the individual stakeholders to provide their own ranking of alternatives, which is then followed by aggregation of these rankings to represent the view of the whole participating group. When carried out in a fully transparent manner this process can possibly lead to negotiations (amongst the stakeholders) towards a consensus on the preferred management alternatives to be implemented. The framework also considers stakeholders' participation in modelling activities (e.g. model validation). Since stakeholders have more knowledge of their local environment, proper assimilation of this knowledge may significantly improve the model results. Moreover, their engagement in modelling-termed Collaborative modelling can be used as a learning process for better understanding of the system in question and some of the introduced measures. The framework can be summarised as consisting of two main stages: (1) Collaborative modelling and (2) Participatory decision making.

The main criteria for selection of a particular NESP framework are case-specific and depend on the environmental characteristics, the type of the management problem and its objectives. Moreover, as part of identifying the NESP frameworks to be used it is important to first asses the case study characteristics and only subsequently design the participatory process. This assessment will also guide the construction and implementation of the NESP.

The NESP frameworks were used to develop and test applications for five case studies with different environmental problems and management objectives. These case studies are the (1) Lakes of Noord Brabant, the Netherlands, (2) Somes Mare catchment, Romania, (3) Danube river (Braila-Isaccea section), Romania, (4) Cranbrook catchment, London, UK and (5) Alster catchment, Hamburg, Germany.

The Noord Brabant case study implemented the *NESP-IKS* framework. It aims to provide up-to-date bathing water quality information about several small lakes located in the study area to various types of users, such as swimmers or surfers.

The *NESP-CP* was implemented for both the Danube River and the Somes Mare catchment. Both case studies are related to flooding issues and aim at improved flood management through awareness raising and information dissemination and sharing among water authorities, professionals and broader stakeholder groups and citizens.

NESP-CDM was applied for the Cranbrook catchment and the Alster catchment case studies, with similar aims of empowering stakeholders in planning and decision making in flood risk management.

For the Noord Brabant case study an integrated web-mobile application was implemented, while for the other four case studies web environments were developed and implemented. In all the case studies these applications were developed and tested in combination with face-to face workshops with the end users / stakeholders. Commonly the NESP deployment was initiated with such workshops, aiming to introduce and demonstrate the NESP applications to the stakeholders; the number of subsequent face to face workshops depended on the type or level of participation. Afterwards, stakeholders were given time to use and test the applications. Finally they evaluated the applications using evaluation forms during the final workshops.

In general the NESPs developed were well appreciated by the users / stakeholders and they clearly recognised the value of using such environments. Water authorities, decision makers and some stakeholders and citizens expressed wishes for extensions of the NESP applications with additional information and development of similar applications for other study areas.

For the development of the applications several technologies have been applied and tested in this research. General Public Licence (GPL) technologies were intensively used for the development. The selection of GPL technologies was critical in building the NESP. Such selection was based on the design of the participatory process and the resources available. More specifically, this research demonstrated that the selection of available GPL technologies must be done carefully following a set of criteria: (1) their applicability within the framework, (2) flexibility and compatibility with other technologies, (3) for the pre-built application components, the general stakeholders should be familiar with their interfaces (e.g. Google maps), (4) the ease of using the technology and (5) the technology should be widely supported by software development community and continually developed.

In general the use of GPL technologies for such platforms is highly feasible. They do provide the desired level of interactivity in the developed components and have the flexibility to be adopted in other case studies. However, it should be also stated that for these kinds of applications programming skills are not sufficient by themselves. When developing platforms for stakeholder participation in water resources or flood risk management interdisciplinary knowledge and skills are needed usually available only in teams of developers with diverse expertise.

The work presented in this dissertation demonstrated that NESP such as web-based and mobile environments have the potential to overcome the hindrances in stakeholder participation in water resources and flood management.

Adrian Delos Santos Almoradie

Delft, the Netherlands

Table of Contents

Chapter 1
General Introduction

This chapter introduces the research on Networked Environments for Stakeholder Participation (NESP) in water resource and flood management. Firstly it presents the research background on stakeholder participation and the use of networked environments for water resources and flood management. Next is an overview of water resources and flood management directives/strategies of European Union (EU) and non-EU countries, followed by a brief introduction to the importance of stakeholder participation. A brief review of several web-based systems for participatory environmental management and their shortfall is also presented. Lastly presented are the objectives and the structure of the thesis.

1.1 Background

Decision making in water resources and flood management (WRFM) is usually implemented through a top-down approach without sufficient involvement of stakeholders. This often leads to blockages and deadlocks in the implementation of the proposed measures. Ideally the decision making in WRFM should be carried out via combining both top-down and bottom-up approaches. Since stakeholders have a better understanding of the real potential and limitations of their local environments, empowering them for participation in planning and decision making is essential for the sustainability of the measures to be adopted (Webler et al., 1995; Abbott and Jonoski, 2001; UN-ESCAP, 2003; White et al., 2010).

Participation in water resources and flood management can be in different forms. It can take place through sharing of information and knowledge or through active collaborative decision making. The nature of this involvement obviously depends on the type of management strategies (e.g. long term planning or event management) and the nature of the problem (e.g. management of watershed, bathing water quality, floods etc...).

The major challenge in stakeholder participation is launching and maintaining the participatory process. The limitation of financial resources, stakeholders' spatial distribution and their interest to participate are some of the possible hindrances in initiating and maintaining the participatory process (WMO, 2006). With the widespread availability and usage of the Internet, researchers and practitioners increasingly try to address these challenges and hindrances by developing and using web-based networked environments.

Networked environments are web-based computer or mobile-aided environments for remote interaction between participating entities such as stakeholders. A networked environment can not only answer the limitation of financial resources and stakeholders' spatial distribution, but this can also provide a more informative and interactive means for participation.

Following the realisation of the potential of using networked environment for stakeholder participation, within the last decade several web-based computer-aided environments have been developed. However, the focus of most such developments was on appropriate structuring and visualisation of decision-making problems, primarily targeting decision makers, without sufficient attention to interactions between decision makers and stakeholders, and even less to interactions among stakeholders themselves. In general there is insufficient research on using networked environments for participation of different types of stakeholders.

In recent years mobile technology has demonstrated even more advantages to reach most of the citizens and potential stakeholders. The advancement of mobile technology and its application development paves a way for its usage for data gathering and information dissemination. In effect, the Internet, World Wide Web, mobile and wireless technologies, present a powerful environment for development and deployment of networked environments as envisaged in this study.

This research presents a generic conceptual framework and specific design and implementation of Networked Environments for Stakeholder Participation (NESP) in WRFM. The NESPs case specific adaptation of the conceptual framework was applied in five real case studies: Noord-Brabant lakes in the Netherlands, two from Romania - the Somes Mare catchment and Danube River Braila-Isaccea section, the Cranbrook catchment in London, United Kingdom, and the Alster catchment in Hamburg, Germany.

The NESPs were developed using the advanced and open source ICT and were implemented for the two types of management strategies, the long-term and event management cases. The NESPs for long term planning supports transfer of knowledge, exchanging of ideas and negotiation to reach a common goal. The NESPs for event management supports awareness raising through information sharing and dissemination.

1.2   Water resources and flood management in EU and non-EU countries

National directives for the management of water resources and floods are important. They provide guidelines and standards for experts, authorities and decision makers in the planning and implementation of management strategies, leading to a better management of water resources.

Most developed countries have implemented legislation or directives for the management of their water resources and floods. In the developing world more and more countries are also developing legislations to better manage their water resources.

The following summarises legislations and guidelines in EU and non-EU countries.

In the EU, the European Commission (EC) established the Water Framework Directive (WFD) and Flood Directive (FD). The WFD, established in 2000, aims at sustainable management of all coastal waters, inland surface waters and groundwater in the European Union and its member states (EC Directive, 2000). Realising that there is a need to establish a directive on the assessment and management of flood risks, in 2007 the EC established the FD (EC Directive, 2007). The FD aims to reduce the negative impact of floods on human health, environment, economic activity and cultural heritage. Moreover, both directives (WFD and FD) encourage the EU and its member states to have management plans that incorporate public information and consultation.

In the United Kingdom (UK), the 2009 Flood Risk regulation and 2010 Flood and Water Management (FWM) act aimed on improving water and flood risk management was established. These regulations encourage policy and decision makers to incorporate short term and medium to long term actions and increase capacities and skills of local authority, citizens and stakeholders (Defra, 2010).

In most developing countries there are no established legislations or official guidelines on flood risk management. The World Bank (WB) initiated the development of guidelines on integrated urban flood risk management (World Bank, 2012). The WB guidelines were based on twelve key principles, some of which are: FRM should consider different scenarios, be designed to cope with changing and uncertain future and FRM should be integrated in urban planning and governance. Of interest here is that according to these guidelines FRM should also encourage multi-stakeholder cooperation and continuous communication to raise awareness and reinforce preparedness.

In summary the directives, regulations, guidelines and practices in water resource and flood management aim not only to properly assess and mitigate impacts of floods, they also recognise that it is crucial to involve the stakeholders and the public in any water-related planning and management.

1.3 Importance of stakeholder participation

The role of people at the local level is crucial in active management of many aspects of water resources. Residents of the local community often have better knowledge on the potential and constraints of their environment. Thus, empowerment of stakeholders who can represent the local people has become an essential objective of many water professionals (Bonn Conference, 2001; Abbott, 2001).

Empowering the stakeholders should be a top-down and bottom-up approach, which means that they should be involved in the planning and management through participation. Lack of stakeholders' awareness, involvement and participation creates problems in disaster management planning.

An example of participation in a long-term planning is by exchanging ideas, knowledge and negotiation to reach a common goal. In event management, participation of stakeholders can be through awareness raising and sharing of information (e.g. information on current water level, flooded area, water quality status).

Stakeholders' participation should not be seen as a burden in water resources and flood management. Instead, it should be treated as an essential part of the management and planning processes, because:

1. It brings together a diverse range of stakeholders to share ideas, knowledge, information, needs and concerns.
2. It helps all stakeholders to be aware of the impending problem and its proposed counter measures.
3. Promotes effective cooperation and understanding between stakeholders.
4. It builds resilience by enabling them to be more knowledgeable about the vulnerable areas, thus providing them adequate information to prepare the community in an event of a water related disaster.
5. It ensures the sustainability of measures adopted.
6. It brings autonomy and flexibility in decision-making and implementation.

More details on the importance and type of participation are presented in Chapter 2, containing the literature review of stakeholder participation in WRFM.

1.4 Towards a Networked Environment for Stakeholder Participation (NESP)

Initiation and maintaining the participatory process is a major challenge for stakeholders' participation in the water resources and flood management. Factors such as spatial distribution, diverse interest of the stakeholders and limited financial resources are some examples that may come as hindrances in maintaining the participatory process. The use of Networked Environment (NE) was hypothesized as a general solution to address these challenges and hindrances.

The advancement of computers and Information and Communications Technology (ICT) led to the World Wide Web (WWW), which has revolutionized the way our society communicates and accesses information. It is foreseen that, with such advancements in ICT, web-based NEs will provide innovative solutions to address the challenges in water-related information sharing, dissemination and stakeholder participation, because of the following advantages:

1. Bringing together all the stakeholders in one environment where they can participate in real time regardless of their spatial distribution
2. Reducing the cost of bringing together the stakeholders
3. Providing innovative tools that could gain and maintain the stakeholders' interest and their long term commitment to the participatory process.

Over the past couple of decades researchers and practitioners have started working on developing the necessary frameworks and applications of NEs for stakeholders' participation. (More extensive literature review on participation in a networked environment will be presented in section 2.6 of chapter 2). Already in the early years of the WWW decision makers and experts have envisioned the use of web-based environments for remote access to decision support systems (DSS) (e.g. Bhargava and Krishna, 1998).

In the field of environmental management, several web-based systems were developed for the management of watershed (e.g Choi et al., 2005) and groundwater (e.g Khelifi et al., 2006; Jonoski, 2002). There are also a number of web-based system that have been developed for flood risk management, however most of these systems fall short in addressing the users' (stakeholders and citizens) needs/requirements and they lack applications usability.

It is evident that many known systems, commonly introduced as decision support systems (DSSs), mainly available as stand-alone applications, still lack important features for stakeholder collaboration. Jonoski (2002) in his research entitled Network Distributed Decision Support System (NDDSS) attempted to utilise the advantages of the web environment for introducing such features.

The NDDSS introduced an approach for promoting stakeholder collaboration using a web-based framework. This web-based framework was envisaged as a DSS deployed on the Internet that allows participation of individual stakeholders including the general public. The framework, although conceived to be quite general, has only been implemented as a prototype for one limited application area of groundwater management.

Further research and development of the NDDSS framework can lead to its possible adaptation to different application areas, which is also one of the main thrusts in this research work. Especially the latest technologies for developing the Internet and the mobile phone applications are still underutilized for developing NE's for stakeholder participation. The latest technologies envisaged in this research will allow for building platforms with higher usability that may provide a new dimension for collaboration.

1.5   Objective of this research

The main objective of this research is to conceptualize, design, implement and test a set of Networked Environments for Stakeholders' Participation (NESP) in water resources and flood management. The NESP were conceptualized to be case specific, given the differences of case studies management and planning and type of participation.

An abstract presentation of a NESP is given in Figure 1.1. Various stakeholders use the NESP in order to access all relevant data (facts) about the system in question, and have access to models results that assist in generating alternatives that can be evaluated with respect to stakeholders' interests / objectives. Data from different sources and models are made available at the back end of the NESP, together with other tools for decision support, such as evaluation tools, negotiation support etc. The primary objective of this research is the case specific design and implementation and evaluation of the generic NESP concept indicated in Figure 1.1.

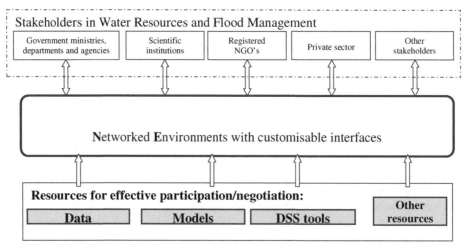

Figure 1.1 Conceptual presentation of a networked environment
Almoradie (2013)

The foremost challenge was to research and construct the framework for a participatory process in a NESP. The specific implementation of this generic framework was developed for different cases using the recent, advanced and cost-effective technologies.

To achieve the main objective several areas were researched. The specific objectives are as follows:

- Investigate existing methods for stakeholder participation, propose improvements and develop new methods for making them usable and effective in NESP.
- Develop the NESPs for several case studies.
- Test the NESPs for specific cases.
- Evaluate the NESPs, tune them and develop recommendations for their use.

The objective was supported by ICT-related technological developments. Web-based GIS, web services and other web technologies were used to enhance usability, visualisation and interactive components of the NESPs.

1.6   Outline of the thesis

This thesis is organised in eight chapters:

Chapter 2 is a review of stakeholder participation in water resources and flood management. It presents an overview of the benefits and potential pitfalls in stakeholder participation and the different types of participation. A brief review of past implementation and experiences in stakeholder participation is also presented (with and without the use of networked environments).

Chapter 3 presents the five case studies. This provides a brief overview of the study areas, the water resources or flooding problems, the management practices and targeted stakeholders.

Chapter 4 presents the NESP conceptual framework, followed by its adaptation to the cases (the application depends on the case study and type of participation).

Chapter 5 presents the information technologies used for NESPs development. A review of existing and latest information technology (web based and mobile) is presented together with criteria for selection of technology for developing the NESPs.

Chapter 6 presents the design of NESPs and software implementation. Firstly it presents case specific conceptual design followed by the final design and software implementation.

Chapter 7 presents the deployment and evaluation of NESPs. It starts with a brief overview of evaluation methods for stakeholder participation. This is then followed by NESP deployment in the case studies (presenting results and implementation experiences and challenges) and stakeholders' evaluation.

Chapter 8 presents the general conclusions of this research and recommendations for NESPs further improvement.

Chapter 2
Stakeholder Participation and its Relevance to Water Resources and Flood Management

This chapter introduces the importance of stakeholder participation in water resources and flood management. A brief overview of the different types of participation is presented followed by a summary of the objectives, benefits and potential pitfalls in stakeholder participation. A review of the lessons learned from previous studies in stakeholder participation and the use of networked environments (NEs) for this purpose is also presented.

2.1 Introduction

Before discussing stakeholder participation, it is important to define what a stakeholder is and who the stakeholders are in water resources and flood management. Stakeholders can be defined as groups or individuals with an interest, or who may be affected or have a stake in the outcome of a certain project. The stakeholders involved in water resources and flood management can be divided in seven groups (1) Government ministries, departments and agencies (2) Affected communities (3) Other basin communities (4) Scientific institutions (5) Registered NGO's (6) Voluntary organizations (7) The private sector (WMO, 2006).

Traditionally, water resources and flood management have been largely implemented through a top-down or top-driven approach. Top-down approach can make decisions very quickly but its implementation may take a long time because some of the stakeholders may oppose the decisions or some of the proposed measures. On the other hand, bottom-up approach would be unrealistic since it practically means that the stakeholders always say what the decision makers should do, while the management would be impossible if there is an absence of the technical support and reality check (UN-ESCAP, 2003). In summary, this mismatch between the top-down and bottom-up approach may often lead to blockages and deadlocks in the implementation of the proposed measures.

Ideally, the decision making in WRFM should be carried out via combining both top-down and bottom-up approaches (Krywkow, 2009). Since stakeholders have a better understanding of the real potential and limitations of their local environments empowering them in planning and decision making is essential for the sustainability of the measures to be adopted (Webler et al., 1995; Fischer, 2000 Abbott and Jonoski, 2001; Reed et al., 2008; White et al., 2010). Moreover, transparency in stakeholder participation that reflects conflicting claims and views may increase trust amongst stakeholders (Richards et al., 2004). Reed (2008) presents a comprehensive review on stakeholder participation for environmental management.

Figure 2.1 shows the different models of participation in decision making.

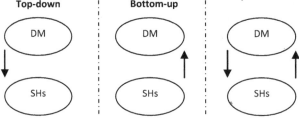

Figure 2.1. Models of participation in decision making
Krywkow (2009) (Modified)

The following section presents the summary of objectives, benefits and potential pitfalls in stakeholder participation.

2.2   Objectives, benefits and potential pitfalls in stakeholder participation

The World Meteorological Organization presented objectives, benefits, challenges and potential pitfalls in stakeholder involvement in integrated flood management as introduced in WMO (2006).

a) Objectives

Stakeholder participation should provide all stakeholders with full opportunities to share their concerns, views, influence the outcomes, build their commitment to the participatory process, ensure implementation of the measures, build resilience and ensure sustainability of plans and decisions.

b) Benefits

Participation of stakeholders should not be treated as an external burden in the river and flood management. It should be treated as beneficial to the decision making process since (1) it brings together a diverse range of stakeholders to share their concerns, views, ideas, information and knowledge, (2) it reduces potential conflicts and promotes effective cooperation, (3) it provides awareness to the stakeholders and the public about the process, (4) it produces better consensual decisions, (5) it builds resilience, (6) it ensures sustainability of measures adopted and (7) it brings flexibility in decision-making and implementation.

c) Challenges and Potential Pitfalls

The challenges in stakeholder participation are in the launching and maintaining the participatory process. These challenges can be attributed to the potential pitfalls that may occur before and during the participatory process. These potential pitfalls are as follows: (1) Participatory processes terminated prematurely or implemented unsatisfactorily lead to disappointments due to high expectations, (2) Participation of all stakeholders may be too costly, (3) Insufficient involvement of real decision makers leads to sub-optimal results, (4) The process is time consuming and requires long-term commitment, (5) Lack of financial and human resources in bringing together the stakeholders over a large geographical area.

The different forms of participation in WRFM are presented in the following section.

2.3   Types of participation

Participation in WRFM can be in different forms. Obviously the form of participation depends on the type of management strategies (long term or event management) and application area (e.g. management of watershed, bathing water quality, floods, etc...).

For the past decades there were several attempts to categorize participation based on the level of engagement. The "ladder of participation" of Arnstein (1969) presents the intensity of participation, ranging from passive dissemination of information to active engagement. Biggs (1989) (one of the most widely used work) categorised the level of participation into contractual, consultative, collaborative and collegiate. Farrignton (1998) developed a simplified version of Biggs (1989) ladder of participation.

One important aspect, common to the different type of participation in environmental management is the integration of scientific and local information or knowledge (Chess et al., 1998; Chase et al., 2004; Fischer and Young, 2007). Proper integration of local information and scientific knowledge may significantly improve management strategies.

Different type of participation will likely be appropriate for different cases, depending on the work objectives and the stakeholders involved (Richards et al., 2004; Tippett et al., 2007).

The following sub-sections briefly discuss the types of participation applied in this research.

2.3.1   Information and knowledge sharing

Participation through information and knowledge sharing are considered as one of the important components for water resources and flood management. The effort of sharing information and knowledge among stakeholders and experts raises their awareness, which will in turn increase local capacity of the communities to deal with water related problems (e.g. floods, bathing water quality). Furthermore, this will enable the stakeholders to understand the areas exposed to water related risk and to equip them with the necessary knowledge on risk assessment and management.

Since the 1960s stakeholder participation has been more about awareness raising and by the 1970s citizens or stakeholders have become more involved in providing data to experts and managers (Pretty, 1995 a,b). However, after more than three decades of awareness raising and collecting information from stakeholders or citizens, these practices are still not common in environmental management. For water resources and flood management, the potential of using information provided by concerned stakeholders such as citizens has already been widely recognised. Developments in which information provided by citizens is used for better understanding and management of environmental systems have been recently described using the term 'citizens science' (Paulos et al., 2008).

2.3.2   Consultative participation

Information and knowledge sharing are applicable to all types of stakeholder participation. However, in some cases this is not sufficient for managing and implementing measures. For the past two decades environmental managers and authorities have seen the growing need for people to more actively participate in the environmental management (Hickey and Mohan, 2005). Consultative participation in environmental management and planning can lead to a more sustainable decisions and measures.

2.3.3   Collaborative decision making

Participation through collaboration is recognised to be one of the most appropriate approaches when designing plans for future water resources or flood management strategies leading to selection of sets of measures and alternatives. Moreover, as presented by Voinov and Bousquet (2010) and Whatmore and Landström (2011), stakeholders gain much more interest to participate when they are to some extent engaged in modelling activities- termed Collaborative Modelling, which are nowadays indispensable in planning processes. The nature of this involvement obviously depends on the available expertise of the stakeholders, but a common level of understanding of the modelling assumptions and capabilities can lead to active involvement in setting-up the modelling objectives and analysis of the results.

2.4   Participatory process

An effective participatory process is important for any type of participation. There are many methods or practice guides regarding participatory processes, for example by Ridder et al. (2005), Elliot et al. (2005), Bousset et al. (2005) and Wates (2000). According to Krykow and Hare (2008) many of the cases miss a systematic or complete guideline on aspects such as social learning, communication, and democratisation.

Based from their experiences and study of participatory process, Krywkow and Hare (2008) proposed to partition the participatory process in four phases, as follows:

1. Preparation
    * Problem analysis, stakeholder analysis, resources analysis, goals analysis, drafting a participatory plan;
2. Publication
    * Presenting to the public and stakeholders the existing problems, objectives of the plan and options for measures and solutions, including the likely impacts on social and physical environment;
3. Dialogue
    * Additional and deeper information provision, knowledge elicitation, education, detecting planning design errors and yet unknown side effects;
4. Response
    * Education, social learning, recruiting volunteers, scenario building, model validation, finding consensus or compromise, adjusting the planning goals

This research followed similar participatory process (especially for collaborative decision making).

2.5   Lessons learned in stakeholder participation

Learning from the results and experiences from the past studies are useful so as to know the obstacles and the promising methods for the participatory process. The lessons learned can provide better approaches for the framework of the NESPs. A brief review and examples of lessons learned from stakeholder participation in water related management and other sectors is presented as follows:

a)   Water resources management

Krywkow (2009) explored a methodological framework for participatory processes in water resources management. In his PhD thesis the participation of stakeholders was implemented and analyzed in five case studies: (1) British Waterways: Stroud canal restoration (2) Glasgow City Council: the regeneration of Ruchill Park (3) POM West-Vlaanderen: constructing a new fresh water basin (4) Water Board of Scieland and the Krimpenerwaard: a new water way (5) The Province of North-Holland: Improving and extending a recreation area. In summary, the most important lessons and experiences learned from these studies are:

- There should be better involvement of stakeholders in the planning of participatory process. Stakeholders with serious concerns must be taken seriously and a dialogue must be pursued.
- Communication with stakeholders must be more efficient
- Better stakeholder analysis to involve earlier the relevant stakeholders. Their early involvement helps to avoid misunderstanding and has the potential to encourage more support for a given project
- Better management of the expectations of stakeholders
- The benefit of a well prepared and organized participatory process was recognized
- The goal oriented approach of participation supports the application of appropriate participatory methods.

b)   Agriculture and irrigation development

Because of the scarcity of water in some countries, a small-scale irrigation scheme that involves stakeholders in its implementation plays an important role in the agriculture and economic development. The lessons learned from the Food and Agricultural Organization (FAO) projects in Zambia, Pakistan and Afghanistan that included stakeholder participation were summarized in the report Flood Management Policy Series (WMO, 2006). These lessons are summarized as follows:

- Community-based activities should be developed gradually
- Tools and methodologies should be flexible enough to meet the changing needs and priorities
- An appropriate strategy for financial and technical support must be developed and sustained beyond the project duration

- It is important to build synergy with existing institutions rather than create parallel structures.

c)   Community-based disaster risk management (CBDRM)

The aim of community-based disaster risk management is to build a culture of safety and ensure sustainable development. Communities were involved in the decision-making and implementation of disaster risk management activities. The lessons learned from the experiences of successful CBDRM approaches were summarized in the report of WMO (2006).

- Synergies are developed when a project makes use of the communities' knowledge and conducts consultative workshops to identify priority action areas.
- Appropriate and low-cost technology acceptable to the community should be transferred to the community
- A transparent management structure through the establishment of a series of advisory committees at the district and community level is useful to create synergy with other administrative structures
- To ensure sustainability it is important to develop linkages with the stakeholders involved on disaster preparation
- Public awareness as a key component of disaster preparedness
- There is a need to set up CBDRM development fund to ensure the continuation of programme activities.

d)   Watershed management (Erosive runoff risks)

To initiate a collective management response to erosive runoff risk at the watershed scale, Souchère et al. (2010) conducted a research that employ a role playing game involving the stakeholders as the players. The game scenario, which is an actual case in Pays de Caux in France, had a recurring problem of erosive runoff. The aim of the study is to use a continuously evolving model to support discussion among stakeholders about the system to be managed and to explore possible future scenarios. Stakeholder participation was done through a role playing game using a participatory approach called companion modelling "ComMod" (Collectif ComMod, 2006). The main idea is to have a series of workshops, for which each stakeholder will receive a map and a list of needed data (e.g the farmers were given a list of crops they can produce together with the gross margins per crop, for the mayor a more detailed map of the watershed was provided with all needed information). The interaction and negotiation takes place in the same venue (altogether) and once each stakeholder has a recommendation a computer operator enters the information in the model and runs it. Once a new watershed runoff map is created from the model, the facilitator gives the new watershed runoff map to the stakeholders. The following are the lessons learned from this study:

- Modelling and simulation can be very useful to accompany a collective learning process

- The role playing game enhanced the participants' awareness of their responsibilities and helped developed their negotiations skills with other stakeholders.
- The ComMod approach combines scientific expertise, political interest, and practical experience. It is viewed as an appropriate way of developing soil protection strategies suitable for practical implementation.

2.6   Information dissemination and participation in a Networked Environment (NE)

The potential use of the networked environments (NE) for communication of decision information, computation, remote and platform-independent access to a DSS has long been envisioned by decision makers and ICT experts (e.g. Bhargava and Krishna, 1998). Within the last decade numerous applications using Networked Environments as a decision support tool have already been developed. These NEs can be standalone or web based environments.

The following case studies used the networked environment for water-related decision making. The NEs was used as support tools for information dissemination or for participation.

2.6.1   Information dissemination in a NE

Choi et al. (2005) developed a Web-based DSS for watershed management. The study explored the relationships between information technology and hydrologic/water quality analysis. The system integrated the Long-term Hydrological Impact assessment model, additional weather data preparation and output analysis tools. It concluded that the DSS integrated with web-GIS, hydrologic models and web application can be useful for decision makers.

Rao et al. (2006) also developed a web-based DSS for managing and planning USDA's Conservation Reserve Program. The study showed its applicability in integrating a modelling component such as SWAT (Soil and Water Assessment Tool) with a web-based GIS-DSS. The ArcIMS GIS platform was used and integrated with the AFIRS algorithm (Automated Feature Information Retrieval System). Web server and Java technology were implemented over an ArcIMS platform to access data and process them in a distributed environment.

Almoradie (2008) conducted a research on the applicability of Hydroinformatics geo-referenced tools and web technologies for flood management. The study focused on the coupling of a flood model and evacuation model and using the web technologies for dissemination. The main feature that was used is a web-based GIS system using the Google maps API of Google Company. The main advantage of using Google maps as a Web-GIS is that it is free to use and is widely used. The modelled flood results (flood extent and flood depth) and evacuation (recommended routes) were disseminated spatially. These applications led to an integrated Web-GIS flood and evacuation information system. The results and documentation of the research demonstrated that the approach opens new ways of disseminating model results, not limited to simple information, but also a full range of hydrodynamic geo-referenced results of the model. In connection to this study, a Web-based GIS using Google maps for the dissemination of water related information was applied in the European FLOODsite project.

2.6.2 Participation in a NE

The following examples of participation in the NE do not explicitly convey whether a top-down or bottom-up approach was used for participation. Based from their context, the level of participation is mainly between decision makers and stakeholders. Thus there is absence of interaction between stakeholders themselves.

Lotov (2003) developed a visualization-based Internet tools that can help lay stakeholders to better understand the issues and study environmental problems in which they can take certain actions. The study discussed how the new democratic paradigm of environmental decision making which involves experts and stakeholders into decision process could be supported by specially prepared web resources. Lotov (2003) concluded that with the use of the IDM (Interactive Decision Maps) technique (which implements an on-line visualization of the trade-off among different objectives using the Pareto frontier) is sufficient enough and convenient for the experts. It was envisioned that IDM techniques can be used by any computer-literate lay stakeholder.

Molkenthin et al. (2001) presented a Web based support system for joint students' analysis of hydro-engineering projects. The results of the study show that the experiment worked well despite the differences in language, nationality, habit, age, culture and educational back ground. The students were able to work collaboratively via the internet. However, the planned objectives were not achieved without problems, mainly due to the large implementation and preparation time required. In conclusion, the study demonstrated the potential and importance of Web based collaboration.

Khelifi et al. (2006) applied a web-based decision support tool for groundwater remediation technologies selection. This decision support tool is a multi-criteria decision making that uses PROMTHEE II algorithm. It was developed to help site owners, investors, local community representatives, environmentalists and regulators to assess the available technologies and to select preferred remedial options.

Promoting individual stakeholder participation by using NDDSS (Jonoski 2002):

This research is more elaborated compared to other literature about stakeholder participation in a NE. It presents an interesting approach in promoting *individual* stakeholder participation, using a web based framework termed Network Distributed Decision Support System (NDDSS). NDDSS was envisaged as a DSS deployed on the internet that allows participation of individual stakeholders, including the general public. Their knowledge and interaction through the system results in emergent development of the decision making process. The general concept of the NDDSSs is divided in two parts, which are the knowledge centre and users' periphery. The transmission of knowledge from a centre (repository of scientific knowledge) to the user's periphery and the social (narrative) knowledge from the user's periphery to the centre is the main task of the system.

To achieve the transmission of knowledge and their mobilization for decision making, the study proposed three basic functional components. These functional components are the (1) Fact engine, (2) Judgement engine and (3) Negotiation and collaboration platform/support. Figure 2.2 shows how the functional components are envisaged to be integrated in the system.

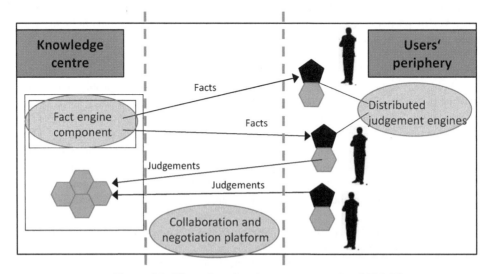

Figure 2.2. Three functional components of an NDDSS
Jonoski (2002)(Modified)

Brief presentation of the three functional components is as follows:

1. The "Fact engine" is the component located in the knowledge centre. This engine gathers, organizes, provides and distributes relevant scientific knowledge as scientific facts *(e.g. Models, data, spatial data, measures, alternatives, strategies),*

2. The "Judgement engine" is composed of components which are located in the users' periphery and are used by a particular user or user group to formulate judgements about the proposed plans or interventions. This engine combines the scientific facts with the beliefs and attitudes of the users/participants in the decision making process,

3. The "Collaboration and negotiation platform" allows for the emergence of commonly accepted and just courses of action with appropriate compensatory arrangements through active engagement in a negotiation process. This component has to aggregate the judgements and evaluations performed by the participants in the system, present the so-called 'social landscape' (positions of the participants

within the negotiation process at hand), explore the options for mutual gains, support alliance building and offer transparency of the negotiation process.

These three functional components were implemented in a prototype NDDSS called Aquavoice (a hypothetical case). The central problem of this hypothetical case is the lowering of groundwater tables due to the installation of a new abstraction wells. In this scenario the community has to decide on the appropriate combination of abstraction wells that will satisfy the water demand while being acceptable to various existing interest. The stakeholders/institutions involved (Local authority, Environmental agency, Private water supply company and the Farmers council) have to negotiate for an acceptable course of actions with compensatory arrangements. In addition to these institutional interests – individual users can join the process by preparing their own judgments and evaluations of the proposed alternatives. The three components used in this hypothetical case are as follows: (1) Fact engine- MIKE SHE[1] model of the area in question, (2) Judgement engine: algorithms from the category of FMADM[2] (Fuzzy Multi Attributes Decision Methods), and (3) Collaboration and negotiation platform: transparent representation of the "social landscape", automatic update of overall response (judgement and evaluations) of the whole community to the proposed alternatives and a chat like interface for direct actual negotiation.

This research concluded that constructing an NDDSSs that is capable of dealing and supporting multiple aspects of participatory decision making process is a very complex task and different decision making problems will involve different tasks and domains, requiring different actual implementation of the proposed NDDSS concept. Consequently, these different implementations may also require different technologies for achieving the required interoperability of the NDDSS components.

This work also concluded that: (1) The primary design of the fact engine should be more towards the use of integrated modelling, which may require integration of different kinds of software components and encapsulation of different kinds of fact-providing knowledge, (2) The design of the judgement engine is one of the most difficult aspects of the NDDSS. The actual designs of this engine is conditioned by our improved understanding of judgement formation processes that needs further development of theoretical concepts and adoption of new kinds of semiotic concepts, (3) The main challenges in the design of the collaboration and negotiation component is the extension of the platforms capabilities with components of "intelligent" support to the users in terms of suggesting options with mutual benefits and possibilities for building coalitions or alliances.

---

[1] MIKE SHE is an integrated modeling framework for simulating all components of the land-phase of hydrologic cycle (http://www.dhigroup.com/Software/WaterResources/MIKESHE.aspx).

[2] **F**uzzy **M**ulti **A**ttributes **D**ecision **M**aking **M**ethods: a Multi Criteria Analysis method that uses fuzzy set theory that addresses a problem in which a number of alternatives in the decision problem is predetermined and finite, and the attributes can be expressed in both quantitative and linguistic terms.

Many of the principles laid in the design of NDDSS were used and extended in the present study.

The role of mobile phone technologies for stakeholder-oriented applications in WRFM:

The use of advanced mobile phone technologies (e.g. smart phones) for water resources and flood management is foreseen to enhance stakeholder participation, initially by personalized information dissemination, and potentially by provision of applications that enable engagement of individual citizens as stakeholders. There are several research studies which try to explore the potential of mobile phone technologies for water-related problems. The following case studies aimed to disseminate relevant information via mobile phone applications:

- Web and mobile technologies in a prototype DSS for major field crops (Antonopoulou et al., 2009)
- Optimisation of Monitoring Networks for Water Systems: Information theory, value of information and public participation (Segura, 2010)
- Bathing water quality information dissemination using smart phones (Jonoski et al. 2012b)

A summary of several research applications for mobile applications in the water domain is provided in (Jonoski et al., 2012a). The above mentioned studies demonstrated the potential of mobile phones technologies for developing applications that can be seen as components of NESPs. Even though the technologies for application development are still quite diverse and in competition (iPhone, Windows Mobile, Google Android), most of them are offering similarly powerful capabilities.

2.7   Concluding remarks

From the lessons learned, WMO (2006) recommended to enhance the stakeholder involvement. To enhance stakeholder involvement is to strengthen the participatory approach through creation of highly aware communities, using better information-sharing techniques (e.g. use of NE's), effective programs or mechanisms. Also worth noting, the capacity-building for stakeholders should be treated as an integral part in enhancing stakeholder involvement. As a mechanism for capacity-building, the following is recommended: training sessions and workshops, use of information technology for remote interactions (e.g. use of NE's), networking for information sharing, internships of key personnel in other organizations, public awareness, learning by doing and by role playing.

The major challenge in stakeholder participation is launching and maintaining the participatory process. The limitation of financial resources, stakeholders' spatial distribution and their interest to participate are some of the possible hindrances in initiating and maintaining the participatory process. With the widespread availability and usage of the Internet researchers and practitioners increasingly try to address these challenges and hindrances by developing and using web-based systems and networked environments. A web based environment can not only answer the limitation of financial resources and stakeholders' spatial distribution, but this can also provide a more informative and interactive means for collaboration.

There are many ideas and technologies that can be used, but it is also evident that most known web-based systems and networked environments are kinds of DSSs, which lack important features for stakeholder participation/collaboration. Especially the latest technologies for developing the Internet and the mobile phone applications are still under utilized for developing NEs for stakeholder participation. While the concept of NDDSS's as NE's supports such stakeholder collaboration, it has only been implemented as a prototype and for one limited application area of groundwater management. The whole approach needs further research and development for its application to river and flood management, in both the fact and the judgment engine components and especially in the collaboration/negotiation platform. Furthermore, water resources and flood management authorities face new challenges on raising stakeholders' awareness by involving them in water resources and flood management. In addition to the challenges related to effective and efficient modes of involvement, the task becomes complex because of the amount of flood related information that needs to be managed for this purpose (collecting, archiving and sharing).

The developed framework and latest technologies used in this research will allow for building platforms with higher usability that provides a new dimension for participation.

Chapter 3
Case Studies Description

"Whenever there is a conflict between human rights and property rights, human rights must prevail."
~Abraham Lincoln

This chapter presents an overview of the five case studies, with their study area, water resources or flooding problems, models used, management practices and involved stakeholders.

The five case studies presented in the subsequent sections are the following:

1. The *Lakes of Noord-Brabant*, the Netherlands. This case study was supported by the Localised Environmental & Health Information Services (LENVIS) research project of the 7th Framework Programme of the European Union (EU) and by the Province of Noord Brabant, The Netherlands.
2. *Danube river (Braila-Isaccea section)*, Romania. This case study and related applications were developed within a research project entitled enviroGRIDS at the Black Sea Catchment, funded by the EU FP7 Research Framework.
3. *Somes Mare catchment*, Romania. Same as the Danube, this was supported by the enviroGRIDS research project at the Black Sea Catchment.
4. *Cranbrook catchment*, London, UK. This case study was supported by a research project entitled Decentralised INtegrated Analysis and Enhancement of Awareness through Collaborative Modelling and management of Flood Risks (DIANE-CM) funded by the second ERANET-CRUE initiative of the European Union.
5. *Alster catchment*, Hamburg, Germany. This was also supported by the DIANE-CM project.

Presented in the case studies are the description of its study area, models used, management plans and the stakeholders. Models were used as a supporting tool for the NESP applications. Some of these models were set-up and developed by the research project partners.

3.1   Noord-Brabant lakes, the Netherlands

Study area

The Province of Noord-Brabant has an area approximately 5,000 km^2. The province is the most live urban area with almost 2.5 million inhabitants. Numerous lakes are used for recreational activities especially during the summer season. Such recreational activities are water sports, bathing and fishing. One of the main concerns for bathing citizens is sometimes the water quality of the lakes drop below the thresholds of health standards. This case study is focused on citizens participation in environmental water quality.

Sixty one lakes were selected for this case study. The province of Noord-Brabant provided the monitored water quality data. Figure 3.1 presents the location of the study area.

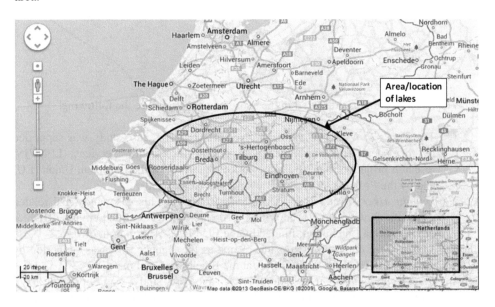

Figure 3.1. Noord-Brabant study area
Almoradie (2013)

A water quality model was set-up for predicting water quality in the largest lake (Binnenschelde) in the Province. The Binnenschelde lake has an average area of 1.78 km^2 and has an average depth of about 1.5 m. It has a maximum depth of 3.5 m. The bathing area located in the eastern part of the lake has an area approximately 1% of the total surface area (Figure 3.2). The lake is connected with its neighbouring lake in Zoommeer through a controlled inlet.

Figure 3.2. Binnenschelde lake
Almoradie (2013)

The water depth in the bathing area varies from 0.6m to 0.3m. The variation in water depth is caused by evaporation, infiltration and the closure of the inlet in the lake of Zoommeer during the period between May and September (Waterchap Brabantse Delta, 2008). Before the start of the summer season, around the beginning of April water from the Zoommeer lake is pumped to the Binnenschelde lake to raise the water level to a maximum of +1.6 m a.m.s.l. (Waterchap Brabantse Delta, 2009). Preventing the water level from exceeding +1.6 a.m.s.l is a spillway on the southern boundary of the lake (Waterchap Brabantse Delta, 2008). The inlet from the Zoommeer is closed during May due to the threat of inflow of blue algae. The blue algae is a form of microcystine that creates a visible layer on the water surface.

Bathing water quality problems

During the summer season, the water quality for some lakes in the Netherlands drops below bathing water quality standards. Algae and contamination from polluted water are the main threats to lakes water quality. Algae concentrations are monitored for some of the major bathing lakes. According to the EU Bathing Water Directive (European Union, 2006), other indicators for bathing water quality alerts are toxic algae (including blue algae), Chorofyl-a, thermo-tolerant coli-group, Escherichia coli (E-coli), and Intestinal enterococcen.

The water quality from connected canals and streams discharging into bathing lakes are seldom monitored and modelled. However, the Binnenschelde lake is frequently monitored and modelled being the largest lake used for bathing.

The main indicator for the lake Binnenschelde bathing water quality alert is on the presence of blue algae. Blue algae is a kind of cyano-bacteria that forms a visible layer on the water surface. Beyond a certain level of blue algae concentration the water maybe toxic for bathers. Blue algae is measured by the concentration of microcystine in cells ml^{-1} or in µgl^{-1}. Binnenschelde also considers other indicators for bathing water quality in accordance to EU Bathing Water Directive.

Figure 3.3 are snapshots of bathing lake in Noord-Brabant and blue algae in fresh water.

(a)   (b)

Figure 3.3. Snaphots of (a) Bathing lake in Noord-Brabant[3] and (b) blue algae in fresh water[4]

Table 3.1 presents the quality standards and alerting practices for microcystine, which was the main modelled parameter.

Table 3.1. Water quality standards and alerts

Water quality parameter	Value for alert	Alert
Microcystine	<=20,000 cells ml^{-1} (4 µgl^{-1})	Pre-alert for water managers
	50,000 cells ml^{-1} (4 µgl^{-1})	Warn not to swim if frequent monitoring is not possible
	100,000 cells ml^{-1} (20 µgl^{-1})	Warn not to swim even though daily checking is possible
	Conc. 200,000 cells ml^{-1} (40 µgl^{-1}) or more	Swimming prohibited

Source: Waterchap Brabantse Delta (2008)

---

[3] http://www.brabant.nl/dossiers/dossiers-op-thema/water/zwemwater.aspx
[4] http://blog.cleanwisconsin.org/index.php/tag/blue-green-algae/

Management plans

Water quality alerts in the Netherlands are provided by regional Water Boards and the National government. Provisions of water quality alerts are through warning signs in the bathing area, radio and television news, websites and TV-text. However, bathers most often missed or ignored these warning signs. Moreover, websites is a passive way of alerting, it depend on bathers if they check the websites before going for a swim. Alert services for bathing water quality using mobile phone SMS are not yet offered. Closing a particular section of the lake for swimming or issuing bad water quality can have significant economic impact for businesses around the lake especially if bathers continue avoiding visiting a lake for several seasons. Bathers may continue avoiding the lake even for cases where poor water quality is restored in short period. Up-to-date information of water quality indicating a good water quality can be beneficial for businesses. In a health safety point of view and for the benefit of public swimmers, providing such information is fully justified.

Citizens participation will be more on information dissemination of lakes monitored water quality data, this is presented later in this thesis.

Stakeholders

The stakeholders identified in this case study are classified into two, professionals and public users. Professional users are the waterboards, national health, environment and water organisations, and consultancy bureaus. Stakeholders who are strongly interested in timely, accurate and continuous provision of water quality information are the beach owners and citizens (bathers).

3.2   Danube river (Braila-Isaccea section), Romania

Study area

The Danube basin is divided into three sections (upper, middle and lower) because of its geographical distribution and differences in hydrological regime. The study area-Braila-Issacea section is located in the lower Danube basin. The study area comprises Galati and the so called Cat's Bend near Grindu (see Figure 3.4). It has two main tributaries named Siret and Prut. During the summer of 2004 and 2005 and in the spring of 2006 a large part of the region was flooded and evacuation was necessary for some parts of the area. The flood events indicated that the Danube floodplain in the Cat's Bend area has insufficient capacity to reduce the flood peak.

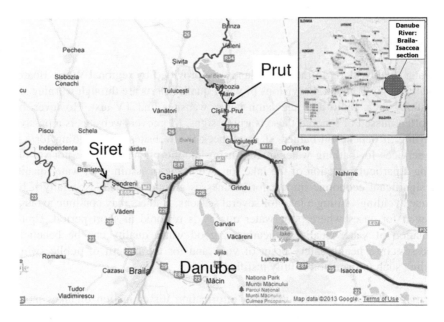

Figure 3.4. Cat's Bend area with locations of rivers Danube, Siret and Prut
Jonoski et al. (2013) (Adopted)

Flooding problems

Galati which is located between Danube's main confluences- Siret in the upstream, and Prut in the downstream has dikes in the lower area. The dikes which were designed to withstand a return period of 1/100 is always in danger of breaching through scouring or internal erosion because of the long duration of flood which can exceed a 100 days.

Flooding in this region especially near Galati is worsened when floods in Danube occurs simultaneously with floods coming from the two main tributaries. Discharge in the upstream of the Prut river will not exceed 600-700 m^3/s because of the controlled reservoir. The Siret river which has no controlled structures has an observed maximum discharge of 4000-4200 m^3/s 65 km upstream.

Management plans

The European Water Framework Directive and Flood Directive is the main instrument for the implementation of the current Danube flood risk management strategies. The WFD and FD encourage development of flood management strategies that does not only protect properties and citizens from floods but also strategies that preserve the ecology of the flood plain. The Danube adopts such management strategy by combining both structural and non-structural measures.

The measures to mitigate flooding in the area are through a series of dikes and using the floodplain in Cat's bend area as a storage basin. Some dikes at certain points along the river were removed to allow the water to flow along the floodplains. Compared to the structural measures such as building dikes, allowing flood water to flow into the floodplain has several advantages, literarily it does not cost anything. For most cases change in spatial planning in the floodplain is needed such as converting agricultural land to natural area.

Moreover, in reference to the Flood Directive, FD encourages inclusive participation of stakeholders and concerned citizen in the planning and management of floods. This has been somehow the missing part in the Danube management strategy. This case study implemented such stakeholder/citizens participation using the developed NESP platform that is focused on flood risk awareness (presented later in this thesis).

Stakeholders

In Romania stakeholders are classified at 3 levels:

1. National level
   - Decision makers that are involved in flood management at government level
   - State agencies playing a role in flood risk management
   - Other agencies at national level

2. Regional/county level
   - Romanian Government Prefecture
   - County Inspectorate for Emergency Situations

3. Local level
   - Local Councils (municipal, town and communal)
   - Local Committees for Emergency Situations
   - Citizens
   - Non – governmental organisations at different levels

In the Danube case study participating stakeholders were identified by the project partners. Although systematic stakeholder analysis was not carried out in this case study, a number of diverse stakeholders were identified and participated in the workshop. The stakeholders were grouped into Government agencies and local councils, Emergency agencies, Companies and businesses, Research institution and General public.

3.3   Somes Mare catchment, Romania

Study area

The Somes Mare catchment (Figure 3.5) is located in the northern part of Romania, and it is part of the larger Somes catchment. The main river- Somes is formed through the confluence of Somes or Somesul Mare (Big Somes) and Somesul Mic (Small

Somes) in Dej city. The Somes river then flows downstream to a confluence that joins with river Tisza (Figure 3.5). The Somes river and Somesul Mic has structures such as cascades of dams, small dams and weirs that protects the area from flooding. While in the Somes Mare river there are no flood mitigating structures and it has been observed that floods are occurring more often.

The case study is focused on the Somes Mare because of its large potential to flooding. The Somes Mare catchment has a total area of 5078 km^2 and the river length is approx. 136 km. It has a complex terrain having an elevation that ranges from 400 m to 2280 m. The annual average discharge of Somes river in Dej is 74.1 m^3/s, to which Somes Mare contributes with approx. 64%.

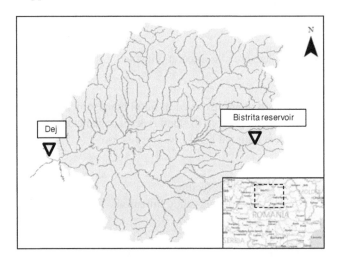

Figure 3.5. Somes Mare catchment
Almoradie (2013) (Adopted)

Flooding problems

A 100 years flood that occurred in the 1970 was the most significant flood in the Dej area. The recorded flood peak in the Somes Mare during that flood event was about 2000 m^3/s. Based from the flood analysis of a 1/ 100 year flood event it shows significant flooding in the Somes Mare. While the analysis for the Somes Mic shows that there are no significant flooding because of the two reservoirs- Tarnita and Fantanele that was constructed in 1974 and 1978 respectively.

Discharge in the Somes Mare is highly influence by snowmelt and when combined with rainfall it causes the most devastating flood. Lately, Somes Mare experienced many occurrences of flash floods. Flooding in the year 2009 was one of the most devastating. The situation in the Somes Mare catchment demonstrates the need for further studies in order to build and implement better flood risk management strategies.

Management plans

The Somes Mare has almost no structural measures to mitigate floods. There is one small reservoir on the upstream tributary- Bistrita however it has very small influence on mitigating floods (Figure 3.5). Responsible authorities and flood managers in the Somes Mare plans to have measures that are in line with the WFD and FD. Non-structural measures such as risk awareness, participatory planning and forecasting and warning are being considered.

Demonstrator application was developed for the Somes Mare (same as with the Danube case study) for flood risk awareness and stakeholder participation.

Stakeholders

Please refer to the previous section (3.2) for the stakeholder classification. The stakeholders in this case study were also identified by the project partners.

3.4   Cranbrook catchment, London, United Kingdom

Study area

The Cranbrook catchment is located in the London borough of Redbridge, UK and it has a small river which is a tributary of the Roding River; the Roding River is a tributary of the Thames. The catchment is mainly an urban area and has a drainage area of approximately 9 km². Figure 3.6 presents the Cranbrook catchment.

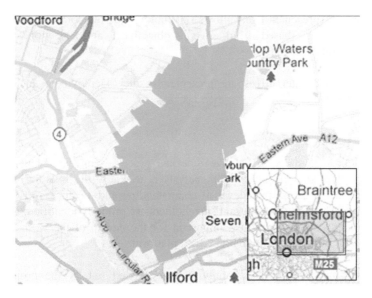

Figure 3.6.Cranbrook catchment.
Almoradie (2013)

Flooding problems

The catchment experiences pluvial (surface) and fluvial (river) flooding. The most recent flooding was reported in February 2009. A forecasting system and warning system with the use of a model has been well established for the fluvial flooding, in contrast little attention has been given to the pluvial flooding. In relevance to the pluvial flooding Pitt (2008) highlighted that this type of flood was responsible for about two thirds of the damages caused by the floods that took place in UK during the summer of 2007. Pluvial flooding has been identified as a missing gap for the flood risk planning and emergency management, thus this type of flooding is the main focus for the Cranbrook catchment case study.

Management plans

One of the main flood risk management guidelines in London borough of Redbridge is the Strategic Flood Risk Assessment (SFRA). The SFRA comprises of the assessment and mapping of different forms or sources of flooding such as groundwater, surface water, sewer, river and tidal flooding. Moreover, it includes the future impact of climate change. The SFRA is an integral part for planning and decision making on future developments in the area; minimising flood damage to property and people (Redbridge, 2010).

The Cranbrook has several structural measures implemented (along the river) for fluvial flooding. Less has been done to address the pluvial flooding. Hence, there is a need to developed flood risk strategy and emergency management for pluvial flooding.

In this case study pluvial flooding was the focus for the development of NESP. The developed NESP (presented later) allows stakeholders to collaborate in the planning of measures/alternatives (based from several objectives) and selection of the best alternatives for implementation.

The objectives and alternatives are presented in the "Deployment and evaluation" chapter- section 7.6.1.

Stakeholders

Stakeholder analysis is one of the important processes in identifying the relevant stakeholders. The aim of stakeholder analysis is to know the relationships between stakeholders and to assess their level of understanding and awareness of flood risk in the study area. Through this, relevant stakeholders are properly identified. Results from this stakeholder analysis were also important for the design of the workshops and to some extent the implementation of the NESP platform.

In this case study stakeholders' analysis were carried out by the project partners (Imperial College of London). Stakeholders were categorized in the following: general public, planners and government organisation, emergency managers and flood management professionals.

3.5   Alster catchment, Hamburg, Germany

Study area

The Alster catchment (Figure 3.7) has an area approximately 578 km^2 and its river has a length of 56 km. The Alster river is a tributary of the river Elbe in North Germany. The lower part of the Alster (Hamburg) is canalised and has been dammed near the centre of Hamburg forming two artificial lakes in the city. The catchment is densely populated in the lower part and less in the upper part. The lower part has a high damage potential from fluvial flooding. Structures were constructed to protect the city from fluvial (river) flooding and influence of high tides.

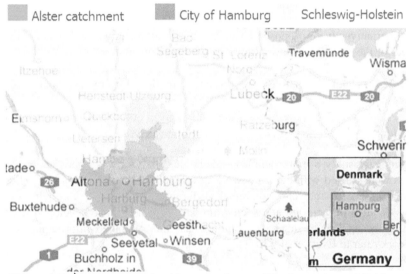

Figure 3.7. Àlster case study: Alster catchment, Hamburg city and Schleswig-Hostein
Almoradie (2013)

Flooding problems

The risk of flooding in the Alster catchment is relatively small. However in the lower part of the catchment there is a high potential to fluvial (river) flooding that can cause enormous damage to properties and people. To protect the city from fluvial flooding and tidal influence structures were built such as controlled weirs and sluices.

During an extreme event there are also possibilities that floods may occur in some parts in the upper part of the catchment. Flooding in the upper part in the catchment can be caused by silted channel, partly connected or clogged natural drainage and topography. These flood risk areas have been identified and was termed "hot spots".

Management plans

The agency for Street, Bridges and Waters (LSBG- abbreviation in Germany) follows' the EU Water Framework and Flood Directive in developing the flood risk management plans. The directives have been implemented in both federal and in state law. The city of Hamburg developed a three-pillar approach for flood protection, which consists of "Preventive flood protection", "Technical flood protection" and "Operational flood protection".

In the Alster there are already existing structural and non-structural measures implemented such as several sluices and clearing the channel from debris. However, as there are still risk in flooding in the lower catchment and localised flooding in the upper catchment, flood authorities and decision makers needs to enhance or further develop its flood protection measures based from the three pillar approach. Same as with the Cranbrook, this case study implemented a NESP collaborative decision making where sets of alternatives were proposed based from several objectives and stakeholders were engaged for the selection of best alternatives for implementation.

The objectives and alternatives are presented in the "Deployment and evaluation" chapter- section 7.7.1.

Stakeholders

In this case study a more detailed stakeholder analysis was done. The stakeholder analyses were carried with the framework developed by the project partners (Leuphana University). The framework combined different methods and through systematic analysis detailed organigram and sociogram were produced. The developed framework can easily be applied to other case studies. For more detailed information about the framework refer to Evers et al. (2011).

As mentioned, stakeholder analysis for this case study was carried out by the project partners (Leuphana University). Five groups of stakeholders were identified: (1) administrative and governmental authorities (at federal and regional level); (2) non-governmental organisations; (3) political bodies; (4) larger business companies; and (5) affected people from the general public. The Agency for Street, Bridges and Waters (LSBG) was engaged as technical partner.

3.6   Concluding remarks

The case studies presented have different environmental problems such as lakes bathing water quality and fluvial or pluvial flooding. Each case study has also different management approaches, the (1) Noord Brabant case study is more focused on the provision of information on lakes bathing water quality, (2) both the Somes Mare and Danube case study is focused more on citizens flood risk awareness leading to their empowerment for participation in flood management and (3) the Alster and Cranbrook case study is focused on collaborative decision making regarding the selection of the most effective and feasible alternatives for implementation. The case studies differences presented a challenge in developing and implementing the NESP.

Moreover, these case studies are in four different countries. Sometimes, language was an obstacle to communicate with the stakeholders. However, different project partners were assisting with the needed translation and communication with the stakeholders concerned.

Chapter 4
NESP Conceptual Frameworks

"To myself I am only a child playing on the beach, while vast oceans of truth lie undiscovered before me."

~*Isaac Newton*

This chapter presents the NESP conceptual frameworks and their adaptation to different cases. The conceptual frameworks follow from the generic framework introduced in Chapter 1, but they are case specific and they depend on the type of participation and case study application. The conceptual frameworks that have been developed are: (1) NESP-IKS for information and knowledge sharing, (2) NESP-CP for consultative participation and (3) NESP-CDM for collaborative decision making.

4.1 Introduction

For an improved management strategy in WRM or FRM, the integration of less structured information (stakeholders' information and knowledge) with structured information (measured and model results) is critically required. Collecting this less structured information through NESP is not that straightforward. The challenge is to gain stakeholders interest to participate and to maintain the participatory process. To gain stakeholders interest, NESP should be appropriately designed, meeting the users' needs and requirements. Moreover, for different WRFM cases the type of participation can be different, given the differences in management strategies, as well as the specific characteristics of case studies (especially with respect to stakeholders' structure). Hence, the implementation of the NESPs for different cases will be different.

Development of case specific frameworks, based on the type of participation, was seen as the most suitable approach in designing the NESP. In the following sections three NESP conceptual frameworks are presented (conceptualised for three different type of participation) and their adaptation to different cases (case studies in this research).

4.2   Conceptual frameworks

The NESP frameworks were conceptualised to address the three different types of participation: (1) Information and knowledge sharing, (2) Consultative participation and (3) Collaborative decision making.

4.2.1   NESP-IKS (Information and Knowledge Sharing)

Citizens or other end users increasingly demand for localised and up-to-date environmental information. This increasing demand presents a new challenge to decision makers, environmental managers and other responsible organisation for the collection, archiving and publishing of environmental data. Cooperation amongst responsible organisations and the use of latest Information and Communication Technologies (ICT's) is seen to provide an effective and efficient collection and publication of environmental information. However for it to be really successful, there is also a need for an active participation of concerned end-users (e.g. stakeholders, citizens). Citizens, stakeholders and other end-users should not only be seen as recipient of information and knowledge, they can also be providers of information. This type of information can be of significant contribution if properly assimilated. End-users information together with official monitoring can keep information up to date. Moreover, observations provided by citizens and stakeholders can reduce the costs of monitoring. More than keeping information up to date, the integration of structured information (official monitored information) with less structured information (stakeholders information) can improve environmental management in different aspects, such as an improve models and forecasts and decision making. Possible applications of this type of participation can be on monitoring of floods, bathing water quality, groundwater supply, outbreaks of water related diseases and similar application areas.

The present framework termed NESP-IKS (**I**nformation and **K**nowledge **S**haring) was conceptualised for proper assimilation of stakeholders' information and knowledge, in order to have more reliable and up to date information. Moreover, the framework offers professionals the possibility of using stakeholders' observations to improve their models and forecast. The conceptual framework (Figure 4.1) has three main components and one optional component: (1) Background information, (2) Information access, (3) Stakeholder participation and (4) Improve model and forecast (*optional*).

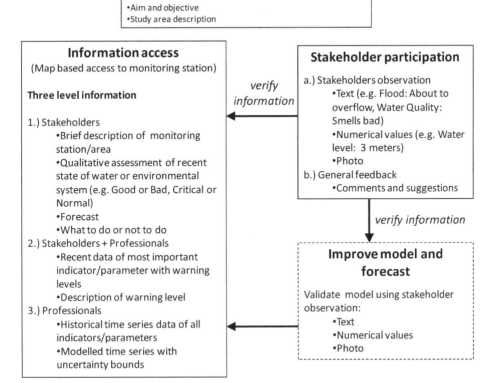

Figure 4.1. NESP-IKS conceptual framework
Almoradie (2013)

NESP-IKS framework provides a three level access to information, conceptualised to target both experts and non-expert stakeholder groups. The three levels of information were intended for (1) Stakeholders (non-experts) - presenting recent information in linguistic and qualitative form, (2) Stakeholder and Professional- presenting useful information for both target groups, and (3) Professionals- presents quantitative information such as historical data and modelled results.

The framework encourages stakeholder to send feedbacks of their observation of the system, this can be through text, numerical values and photos. The feedbacks such as stakeholders' observation can be an added value for experts (to validate their models) and it helps to keep information of the system up to date. However, before usage the information should be verified.

The latest in ICT provides possibilities to develop effective, efficient and innovative applications to disseminate localised environmental information to the wider audience such as citizens. Information and knowledge sharing can be through computer and mobile phone web based applications or an integrated web–mobile applications.

Applications can be developed either as a web browser based or a separate interface application.

4.2.2   NESP-CP (Consultative Participation)

Promoting more inclusive participatory approach for stakeholders exposed to water related risk can influence existing practices in water resources or flood risk management. Inclusive participation such as consultation with stakeholders is encouraged in the development and implementation of management strategies. Developing management strategies based on shared interests and objectives among different stakeholders and the concerned citizens leads to a more sustainable decisions. Furthermore, the knowledge of water authorities and experts combined with knowledge available in the affected communities will significantly improve the assessment and management of water resources or floods.

The framework termed NESP-CP (Consultative Participation) (shown in Figure 4.2) was conceptualised considering the users' needs and requirements. It was conceptualised to be simple yet informative and to have an effective and efficient consultative participation. The conceptual framework has three main components: (1) Risk awareness, (2) Information access and (3) Stakeholder participation.

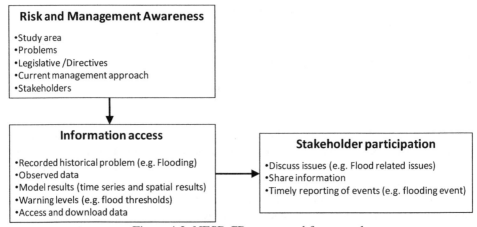

Figure 4.2. NESP-CP conceptual framework
Almoradie (2013)

Most likely general stakeholders such as citizens have little knowledge on the risk and management of their local environment. The "Risk and Management awareness" component is introduced to equip them with the necessary background knowledge on water resources management (WRM) or flood risk management (FRM) legislation, current practices and major stakeholders. This component also provides information on the objective of the NESP platform, the characteristics of the study area and general information on water resources related problems of their local environment.

The "Information access" component is intended to raise the stakeholders awareness on local problems (e.g. flooding). This can be through presentation of historical information of the problem, observed data and model results (time series and maps). The addition of warning levels (e.g. flood thresholds) can be an added value in the presentation of observed and model results. This can be useful for stakeholder participation, providing them some sort of baseline if an event is critical to take appropriate action.

Since stakeholders have a better understanding and knowledge about their local environment their contribution will enhance WRM or FRM. The Stakeholders' participation component provides the stakeholders with opportunities to discuss issues, share information and timely report critical events.

4.2.3   NESP-CDM (Collaborative Decision Making)

A more interesting type of participation is through collaborative decision making. Collaborative participation is recognised to be one of the most appropriate approaches when designing plans for future water resources or flood management strategies leading to selection of sets of measures and alternatives. Moreover, stakeholders gain much more interest to participate when they are engaged in collaborative decision making (e.g. collaborative modelling, selection of best alternatives)- Collaborative modelling engage stakeholders in an iterative and interactive process to understand the flooding problems or how measures influence the flooding patterns through the use of models and other communication tools (Evers et al., 2011).

The framework NESP-CDM (Collaborative Decision Making) presented in Figure 4.3 allows for stakeholder collaboration in identifying realistic management alternatives that address relevant management objectives and subsequent ranking of these alternatives in order of preference by individual stakeholders as well as by the whole participating group. Given that these activities are carried out in a fully transparent manner they can possibly lead to common agreements about preferred management alternatives to be implemented. Moreover, stakeholders' engagement in modelling activities (e.g. model validation) was considered in this framework. The proper assimilation of local knowledge can significantly improve model results. Moreover stakeholder engagement will also help them understand the system and some of the measures introduced.

The collaborative framework consists of two main stages: (1) Collaborative modelling and (2) Participatory decision making.

The first stage is a process in identifying management objectives and alternatives, and testing of alternatives within external scenarios. An initial set of objectives, alternatives and scenarios is proposed by the participating experts (including modellers), but these are further refined through a discussion with the involved stakeholders. This process leads to development of a common shared understanding of the system in question, existing risks, and alternatives for its reduction in future. This stage is structured in four steps: (1) System definition: information about the study area, identification of problems (e.g. floods) and legislation (2) Current risk situation: risk assessment and management, and objectives identification (3) External scenarios: risk assessment and scenario testing (4) Proposal and refinement of the alternatives. This first stage of collaboration already provides an opportunity for learning about the values / interest of other stakeholders who are involved in the process. This is important as such learning may potentially lead to identification of new alternatives and redefining of the objectives that were originally proposed by the experts. It was envisioned (and subsequently implemented) that the interactions in this first stage of collaboration are most productive if implemented through a series of workshops combined with usage of a web platform where all necessary information is provided and updated. This provides more flexibility for the stakeholders and for the modellers/experts who act as mediators between the encapsulated expert knowledge in various models and the stakeholders. This occurs intensively during the workshops. In the periods between workshops the web platform serves as a medium through which modelling experts are made available to support the stakeholders (e.g. providing additional information, additional model simulations etc.).

The second stage of collaboration is participatory decision making. Parts of the framework of Jonoski (2002) were adopted and enhanced for this stage of collaboration. Participation is carried out via a web-based decision making environment. The feedback from the first part of collaboration (identification of objectives and alternatives) provides the basis for the final design of this web-based decision making environment. The process for this collaboration is in three steps: (1) Individual profile - this step provides an environment for each stakeholder to evaluate and rank alternatives based on their own preference regarding the importance of the identified objectives; (2) Group profile - a step that aggregates the results from all individual stakeholders (resulting in the so-called group ranking) and provides transparent representation of the individual positions versus the group; (3) Collaboration and negotiation - a step in which the stakeholders negotiate towards a consensus, with possible adjustments of individual rankings resulting from such negotiations.

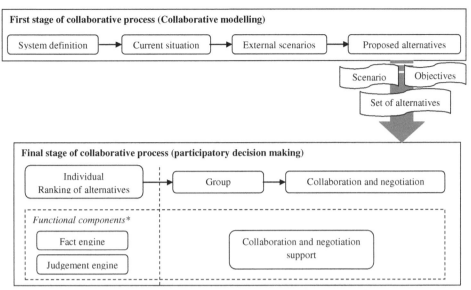

Figure 4.3. NESP-CDM conceptual framework
Almoradie et al. (2013) (Adopted)

The three steps in the second stage of collaboration process make use of the corresponding components introduced by Jonoski (2002): the Fact and Judgement engine for the Individual profile and the Collaboration and negotiation support for the Group profile and Collaboration-Negotiation. They are briefly described as follows:

a.   Fact engine

This component gathers, organizes, provides and distributes relevant scientific knowledge as scientific facts (e.g. Models, data (including spatial data), resulting from various measures, alternatives and strategies under different scenarios. Stakeholders' access such information primarily from pre-prepared model results using a map-based environment and supporting tables. This component guides the stakeholders in their evaluation of the alternatives.

b.   Judgement engine

This component is used by the stakeholders to formulate judgements about the proposed plans or interventions. The engine combines the scientific facts with the beliefs and attitudes of the users/participants in the decision making process. Such combination occurs via appropriate (preferably customisable) user interfaces that enable the stakeholders to clearly formulate their judgements in a way that will also be transparent to other participants. The individual ranking of the proposed alternatives is a final result of the judgement for which Multi Criteria Analysis methods can be used. In this case the method known as Technique for Order Preference by Similarity to Ideal Solution (TOPSIS) by Hwang and Yoon (1981) was chosen. TOPSIS is a Multi Attribute Decision Method (MADM) that uses the notion of ideal points (or ideal solutions) and determines the ranking of alternatives on the basis of their distances from ideal points. Simonovic (2009) pointed out that ideal-point approaches offer advantages for their application in decision-making processes because of their simplicity, transparency and easy adaptation in comparison to other methods.

c. Collaboration and negotiation support

Collaboration and negotiation support is a component aimed to support the emergence of commonly accepted and just courses of action with appropriate compensatory arrangements through active engagement in a negotiation process. This component has to aggregate the judgements and evaluations performed by the individual participants, present the positions of the participants within the negotiation process at hand (the so-called 'social landscape'), allow the exploration of options for mutual gains, support alliance building and offer transparency of the negotiation process. All these activities need to be supported by suitable visualisation components.

4.3   Adaptation of the framework to different cases

The NESP frameworks presented were conceptualised for water resources and flood management. The main criterion for selection of a NESP framework is the case study's environmental and management problem and its objectives. It is important to first asses the case study characteristics before conceptualising the type of participation and its participatory process. The assessment of the case study is important since this provides an overview of the resources and its limitations which is the basis for the design of the participatory process. Such resources are observed data, models, involved stakeholders and their interests, available alternatives, etc.

Once a type of participation has been identified the next step is to design the networked environment. The presented conceptual frameworks provide an innovative approach in designing the NESP.

The presented case studies in chapter 3 have different environmental problems and management. Selection of NESP framework for each case study was based from the assessment and identification of the type of participation. The selected NESP framework for each case studies are summarised in the following table.

Table 4.1. Case studies NESP framework

Case study	NESP framework
Noord Brabant lakes (Netherlands)	NESP-IKS
Danube river (Romania)	NESP-CP
Somes Mare catchment (Romania)	NESP-CP
Cranbrook catchment (UK)	NESP-CDM
Alster catchment (Germany)	NESP-CDM

The Noord Brabant lakes case study, focused on providing up-to-date bathing water quality information for bathers, implemented the NESP-IKS framework. The case study so far have not used bathers' observation in updating the actual bathing water quality information. Integrating information provided by the bathing citizens with official monitoring can provide more up to date water quality information. There is an increasing demand by decision makers to have water quality models as a supporting tool in managing the lakes. Most water quality models come with a high degree of uncertainty because of insufficient data, improving the model results can be achieved with a denser monitoring network and more frequent sampling. However, increasing the monitoring network also increases the cost of monitoring. Hence, most likely experts and decision makers will not invest in a denser network. Another way to improve model accuracy is to assimilate water quality information from citizens. Once properly assimilated this can be an added value for decision making. Presented later, web-based and mobile applications were developed in this case study for such kind of participation.

The NESP-CP was implemented for both the Danube River and Somes Mare catchment case studies (both in Romania). Both case studies are related to issues of flooding. The NESP-CP implementations aim at improved flood management through awareness raising and information dissemination and sharing among water authorities, professionals and broader stakeholder groups and citizens. This objective has been introduced in recognition of the added value of inclusive flood risk management approaches, in which water authorities and experts combine their knowledge and assessment about flood risks with knowledge available in the affected communities. This approach can lead to increased flood risk awareness and consequently better preparedness in the communities. In the long term, it can contribute to developing flood risk strategies based on shared interests and objectives among different stakeholders and the concerned citizens.

The Cranbrook catchment (London, UK) and the Alster catchment (Hamburg, Germany) case studies both aim to empower stakeholders in planning and decision making in flood risk management. Participation through collaborative decision making is recognised to be one of the most appropriate approaches when designing plans for future FRM strategies leading to selection of sets of measures and alternatives. Both case studies adopted the NESP-CDM framework, implementing the two main participatory processes: (1) collaborative modelling process: this is understood as a joint learning process involving FRM experts, authorities and stakeholders, critically supported by models and modelling results. In this process all participating actors are engaged in developing FRM alternatives, objectives and critical flooding scenarios that are further analysed by suitable models, while the actual modelling is still primary responsibility of the modelling experts. (2) Participatory decision making: the end outcome of this process is to come-up with agreed appropriate sets of decisions regarding flood management. The feedback from the first part of collaboration (identification of objectives and alternatives) provides the basis for the final decision making environment. In this process, stakeholders have to evaluate and rank alternatives based on their own preference regarding the importance of the identified objectives. Individual ranking results are then aggregated as group ranking, providing transparent representation of the individual positions versus the group. The final process is for stakeholders to negotiate towards a consensus, with possible adjustments of individual rankings resulting from such negotiations.

4.4   Concluding remarks

The NESP frameworks developed were designed to address three different types of participation in water resources and flood risk management, namely (1) information and knowledge sharing, (2) consultative participation and (3) collaborative decision making.

Before the selection of the appropriate NESP framework to be used it is important to first asses the case study characteristics and then to conceptualise the participatory process.

Although the NESP were applied for bathing water quality and flood management, the frameworks are seen to be applicable to other environmental management problems as well.

Chapter 5
NESP Information Technologies

"We live in a society exquisitely dependent on science and technology, in which hardly anyone knows anything about science and technology".

~*Carl Sagan*

This chapter presents the information technologies available for the NESPs development. It starts with an introduction, followed by a review of the existing and latest web-mobile information technologies, with brief overview of the standards for publishing spatial data and time series over the web. Criteria for selection of appropriate technologies for NESP development are then presented.

5.1   Introduction

The recent advancement of Information and Communications Technology (ICT) is foreseen to provide innovative solutions for the development of the NESP. Since the beginning of the computer and the Internet era, the World Wide Web has been increasingly used in our societies as a technology to access sources of information and communication among the most common individuals. In recent years mobile technology has demonstrated even more advantages for reaching and empowering the citizens and potential stakeholders. The advancement of mobile technology and its application development paves a way for its usage for data gathering, remote execution of models and information dissemination. In effect, the Internet, World Wide Web, mobile and wireless technologies, together with distributed computing resources, present a powerful environment for development and deployment of NESPs as envisaged in this study.

Among the issues related to the NESP development is also the need of the water community for tools and methods for delivering applications that can be interoperable and that can share spatial and time series data over the web. Current trends in archiving, sharing, accessing and managing spatial data are geared towards the use of Spatial Data Infrastructures (SDIs). SDIs are designed for interoperability, allowing publishing of data from any spatial data source using open standards. For sharing, accessing and managing time series data, the Open Geospatial Consortium (OGC) accepted the

WaterML 2.0 schema as an encoding standard for publishing time series of hydrological observation data. These standards are presented more in detail in the succeeding section.

While many commercially available technologies for developing the NESPs is highly usable, customisable and interactive applications do exist, in this case it was decided to make use of General Public License (GPL) internet technologies. Given the recent advances in recent GPL technologies it is now possible to develop highly interactive interfaces that may provide the needed support for the intended application.

The following section presents a review of GPL technologies used for developing the NESPs.

5.2   Review of technologies for NESP

This section presents a review of web based and mobile technology used for developing the NESPs, followed by an overview of Spatial Data Infrastructure (SDI), OGC standards, and particularly WaterML as a standard for time series data.

5.2.1   Web based technologies

Following the realisation of the potential of using web-based environments for stakeholder participation, several web based applications have been developed in the past. However many known systems (commonly introduced as decision support system (DSS) still lack important features for stakeholder collaboration. Furthermore, most of these web-based environments were developed using commercial technologies.

With the recent advancement of ICT and GPL internet technologies, a web-based environment such as the NESP can be developed with similar interactivity and functionalities as with commercial applications. These GPL web technologies are summarized below.

Apache server and MySQL RDBMS- which is under a GPL is a widely used and tested server and database technology. Development of the web interface and its needed interactivity and functionalities can be possible with the use of standard technologies such as HTML, CSS, Javascript, and AJAX- for the client side and PHP scripts and Keyhole mark-up language (KML)- for the server side. Pre-built GPL interface applications that are readily available to be used can be combined in the development of the platform. These pre-built component applications are the Content Management System (CMS), flash animated images and videos, forum, chat module, Google visualisation, Google maps, Bing Maps by Microsoft and OpenStreet Maps (base maps) and the Openlayers (publish and manipulate geospatial data).

Google maps alone, or in combination with Openlayers can be used to present spatial information. Spatial information is provided in the form of point or polygon features. Point features can be used to present locations of hydrometric stations, infrastructure and hydraulic structures. While polygon features or raster image can be used to present properties and flood modelling results (extent and depth). Spatial data may come as a raw data or as results from different models. However, these spatial data may need GIS tools for pre-processing before publishing, such tools are the ArcGIS 9.1, Quantum GIS and Google earth.

PHP scripts with MySQL can be used for dynamic data manipulation. Such data manipulation are log-in, submitting inputs and evaluation results, storage and display of stakeholders' individual profiles and pulling out from the database the scores of the alternatives for use in the group profile.

The use of spatial data infrastructure and its technology are presented separately (see section 5.2.3).

5.2.2   Mobile technologies

The use of advanced mobile phone technologies (e.g. smart phones) for water resources and flood management is foreseen to enhance stakeholder participation, initially by personalized information dissemination, and potentially by provision of applications that enable engagement of individual citizens as stakeholders.

A promising mobile phone operating system is the Google Android platform. Google Android platform is an open source platform for developing mobile phone applications. Led by Google, the platform was developed by Open Handset Alliance which is composed of 65 technology and mobile companies. The first version of the Google Android platform appeared in 2007. In the end of 2008 the first device (G1 phone by HTC) in the market with a Google Android platform appeared.

In the beginning the platform only has a few percent of the smart phone market. However, in recent years the Google Andoid platform has seen an unprecedented growth in the market share. Based from recent surveys the market share of the Android platform in the USA, India and other countries is now more than 50% (~90% in India). As predicted it has surpassed the market share of the iPhone OS. The iPhone Operating System (OS) is the leading smartphone OS in terms of diversity and sophisticated applications. The primary reason the Android platform surpassed the iPhone OS in the market share is because it is an open source platform. This allows other mobile phone producers to make use of the Android platform as the OS. Samsung, HTC and Motorola are among the big manufacturers who have decided to adopt this platform for their devices.

The Google Android platform is a software stack composed of operating system, middle ware and core applications. The OS is a Linux version, while applications are coded in Java programming language. Android libraries and Java core libraries are included in the Android altogether. The Dalvik virtual machine (an optimised version of Java virtual machine on mobile devices) runs the Java applications. An Application

Programming Interface (API) for developers to develop their own application that can be deployed on the mobile device is provided in the Android platform.

Android as an open source platform and Java as its main programming language provides flexibility and ease of developing applications. Hence community developers can offer consumer applications that are for free or at a cost. The architecture of Android operating system is presented in Figure 5.1.

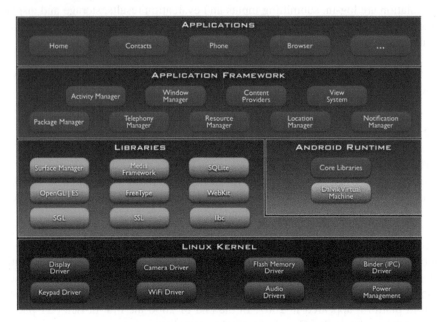

Figure 5.1. Android operating system architecture[5]

Development of application is made possible through the use of Android plug-in for Eclipse environment. Eclipse is an application development environment for Java programmers. Testing of applications can be directly through mobile devices or through an Android emulator. Final versions of applications can easily be deployed on actual mobile devices from Eclipse. Android emulator is shown in Figure 5.2.

[5] http://www.android-app-market.com/android-architecture.html

Figure 5.2. Android Emulator[6]

Android features several applications and technology that are useful for developing NESP applications. Such features are its flexible application framework, integrated open source browser, 2D and 3D graphics library, SQLite data base system, media support for common audio, video, and still image formats (MPEG4, H.264, MP3, AAC, AMR, JPG, PNG, GIF), GSM Telephony (hardware dependent), bluetooth, EDGE, 3G, and WiFi (hardware dependent), camera, GPS, compass, and accelerometer (hardware dependent).

Layar

An interesting third party application that can be used for the NESP is the "augmented reality browser" named Layar. The concept of Layar is to visualise or overlay information in the natural landscape via the mobile phone camera. Layar uses the phone compass and GPS to calculate distance and direction of geo-referenced points coming from a database. The distance from users location and the attributes of the overlaid points (name, data, other information) are presented. The points overlaid in the landscape are customizable (image, size and colours). Figure 5.3 is a sample snapshot of the first version of Layar application developed for the Noord-Brabant NESP and the Layar architecture.

---

[6] http://developer.android.com/tools/devices/emulator.html

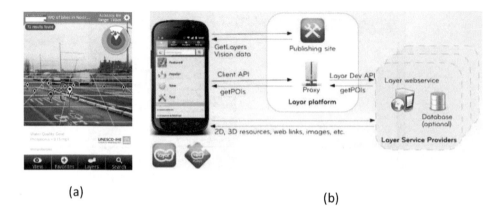

(a)

(b)

Figure 5.3. (a) Sample Layar application *Jonoski et al., (2013)*
and (b) Layar architecture[7]

Given all the advantages and features of Android presented above, the application
presented in this article was developed with this platform.

5.2.3   Spatial Data Infrastructure and Water Mark-up Language (WaterML) 2.0

Spatial data infrastructures are designed for interoperability, allowing publishing of
data from any spatial data source using open standards. Efficient data sharing within the
SDI is enabled by using latest standards for spatial data, such as the Open Geospatial
Consortium (OGC) standards and the standards defined by the EU INSPIRE
framework.

In several NESPs, OGC standards such as Web Mapping Services (WMS) and Web
Feature Services (WFS) have been used for delivering the spatial data. WMS is a
standard for generating maps on the web. Using WMS makes it possible for clients to
overlay maps from several different sources. An important distinction must be made
between WMS and Web Feature Service. WFS is a standard for getting raw vector data
- the 'source code' of the map - over the web. Using compliant WFS makes it possible
for clients to query the data structure and the actual data. Advanced WFS operations
also enable editing and locking of the data.

SDI technology such as GeoServer provides support for Open Geospatial Consortium
(OGC) WMS and WFS standards. GeoServer is an open source software server written
in Java that allows users to share and edit geospatial data. Designed for interoperability,
it publishes data from any major spatial data source using open standards. Being a
community-driven project, GeoServer is developed, tested, and supported by a diverse
group of individuals and organizations from around the world.

---

[7] http://www.layar.com/documentation/browser/layar-platform-overview/

For temporal data, WaterML 2.0 is promoted by OGC as a standard for encoding in-situ hydrological observations data. Water Markup Language (WaterML 2.0) is a standardized Extensible Markup Language (XML) designed for publishing hydro-meteorological observations and measurements time series over the internet via web services.

The known system that publishes time series in WaterML via web services (called WaterOneFlow) is the CUAHSI-Hydrologic Information System . This system has been tested and is used by several US environmental agencies. Commercial products are used together with the CUAHSI-HIS, such commercial products are the Microsoft Windows 2008 Server, Microsoft ASP.NET 2.0, Microsoft SQL Server 2008, ESRI ArcGIS 9.3.1 Desktop, ESRI ArcGIS Server 9.3.1 for .Net (Enterprise Advanced).

A team from Commonwealth Scientific and Industrial Research Organisation (CSIRO) in Australia developed a framework and methods that implements' the WaterML 2.0 schema using the GeoServer Web Feature Services (WFS) with the Postgresql database system. CSIRO developed the GeoServer-WaterML 2.0 to publish water storage time series information from the Australian Bureau of Meteorology (BOM). The use of GeoServer for WaterML 2.0 provides more interoperability.

5.2.4   Other technologies

Standard web based technologies have been well established. Interesting web technologies that can be used for NESP for storing and computing are the Grids or Clouds. The computing infrastructure termed "Grid" is somehow similar with the Cloud concept. The main difference is the Grid can be built locally for the organisations' use and Clouds are usually available commercially.

Foster et al. (2002) & Baker et al. (2002) stated that the Grid infrastructure can provide benefits to many applications, including collaborative engineering, data exploration, high-throughput computing, and distributed supercomputing. The Grid can be viewed as an integrated computational and collaborative environment. Grid computing with the use of applications enables users to effortlessly take advantage of the vast amounts and types of computational resources. However, it has been a very difficult task in adapting applications to run on a Grid given the heterogeneous and complex nature inherent in the Grid. A number of gridification approaches have been proposed to provide appropriate method to gridify applications (Mateos et al., 2008).

Sultan (2009) discussed that cloud computing is an emerging computing paradigm that could provide opportunities for delivering computing services in a variety of ways that have not been experienced before. It has huge potential in reducing the IT investments in an organization, and even completely freeing it from the expense and time of having to install and maintain applications locally. The main drivers of cloud computing are on its economics, simplification, and convenience in delivering computing-related services.

The use of cloud and grids has been applied so far in education and business sectors and to an extent to some environmental models analysis. Although the use of clouds and grids were not implemented in this research, clouds and grids computing potential for water resources and flood management may offer advantages in establishing a cost effective, reliable and efficient model applications (e.g. for flood forecasting and warning) that can be part of the NESP.

Regarding web technologies, it needs to be mentioned that HTML 5, although still in a relatively early phase, offers new possibilities for cross-browser applications for web-computer and mobile devices.

For mobile technology, the most interesting third party application that can be applied in water resources or flood management is the Layar augmented reality browser (previously presented). So far the most developed functionality in Layar is the point-based layer. However, there are now major developments in using 2D and 3D layers as an overlay for the mobile phones camera. It will be interesting to apply these multi-dimensional layers in water resources and flood management (e.g. presentation of historical or predicted floods in 2D or 3D).

5.3   Criteria for selection of technology

For enabling effective participation in a NE there should be a technology for visualization such as interactive maps and charts, real time communication, computational capabilities, interactivity and mobile technology as a support tool (for some NESPs). Since the NESPs are platforms for participation, their design and construction should be developed carefully. This process has to follow the designed participatory process, the designed functional components and accordingly to select appropriate technologies. The use of commercial technologies is always an option. However, nowadays the use of GPL technologies can produce comparable functionalities and output.

Selection of available GPL technologies must be done carefully following a set of criteria: (1) their applicability within the framework, (2) flexibility and compatibility with other technologies, (3) regarding the pre-built application components, the general stakeholders should be familiar with their interfaces (e.g. Google maps), (3) the ease of using the technology and (4) the technology should be widely supported by international community and continually developed.

5.4   Concluding remarks

In developing the NESPs to have an interactive, customizable, interoperable and flexible applications technologies should be carefully selected following the set of criteria. Combination of pre-built applications with some newly built components are usually needed to achieve the needed functionalities.

In general the use of GPL technologies for the development of NESP is highly feasible. As presented later in this thesis, they do provide the desired level of interactivity and have the flexibility to be adopted in other case studies.

In this research computer web based technologies have been implemented for all case studies. For the Noord Brabant case study, a web computer-mobile demonstrator application has been implemented. The following chapter presents the NESP design and software implementation.

Chapter 6
Design of NESP and Software Implementation

"The only way to do great work is to love what you do. If you haven't found it yet, keep looking. Don't settle. As with all matters of the heart, you'll know when you've found it."

~Steve Jobs

This chapter introduces the conceptual and final design and implementation of the NESP applications for the identified case studies. The NESP platforms were designed on the basis of the frameworks introduced in Chapter 4.

6.1 Introduction

In Chapter 4 we presented three novel NESP conceptual frameworks addressing the different type of participation in water resources and flood risk management. In summary, the frameworks are: (1) NESP-IKS - this framework provides a basis in designing a platform for information and knowledge sharing. It has two main components, the information access and stakeholder participation. Participation is through feedback of stakeholders' observation of the system. (2) NESP-CP - is envisaged to be applied in designing a platform for consultative participation. It has also two main components, information access and stakeholder participation. Stakeholder participation for NESP-CP is more than information and knowledge sharing; it also provides options for stakeholders and experts to discuss environmental problems, risk assessment and management strategies. Lastly the (3) NESP-CDM - provides a basis in designing a platform for collaborative decision making. It has two main stages, collaborative modelling and participatory decision making. The first stage is a process in identifying management objectives and alternatives, and testing of alternatives within external scenarios with the use of models. The main aim for the second stage is for stakeholders to negotiate towards a consensus for a possible alternative for implementation. The process for this collaboration is in three steps, through individual assessment and ranking of alternatives, group ranking and negotiation.

As mentioned the NESP platforms presented in the following sections were designed on the basis of the framework introduced. Initially, however, the general design of the front end of the similar NESP platforms will be presented. This is common for similar platforms and serves as a basis for designing the actual web-based user interfaces. Furthermore, most of the frameworks made use of models as a support tool for dissemination, participation and decision making.

6.2   NESP-IKS: Noord Brabant lakes

This section presents the conceptual design, the modelling support, final design and implementation of the NESP for the Noord Brabant lakes case study. Integrated web-mobile application was developed for information sharing on bathing water quality.

6.2.1   Generic conceptual and final design

Generic conceptual design

This case study was designed based upon NESP-IKS framework. It deals with the dissemination of measured water quality information of several small lakes through an integrated web-mobile application. Measured water quality and quantity data of these lakes has been supplied by the Province of Noord Brabant. This application is intended to be used by the general population for swimming, bathing, surfing, etc. However, it was also conceptualised to present relevant information useful for decision makers or specialists. The following figure presents the generic conceptual design.

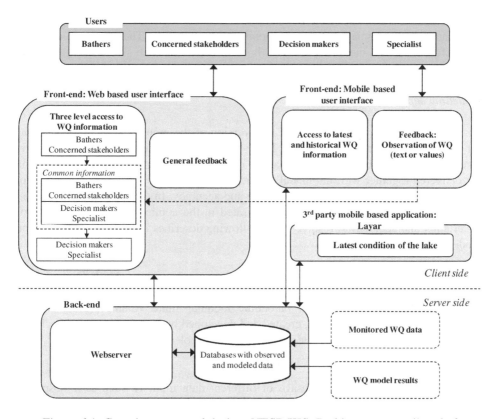

Figure 6.1. Generic conceptual design- NESP-IKS: Bathing water quality platform
for Noord-Brabant lakes
Almoradie (2013)

For users to understand the objective of the platform and to easily navigate and locate
the lakes, the web and mobile platforms has been designed to be straight forward,
simple and with clear layout.

The web platforms' access to information is in three levels. The first level is designed to
provide useful information and understandable alert levels for the non-experts (bathers
or concerned stakeholders). Using the website users can check the current condition of
the lake(s) they want to visit. The second level presents information that is useful for
both non-experts and experts (decision makers and specialist), such as presentation of
important parameters with warning levels and its explanation. The third level is
intended for experts to access information of all historical or modelled data.

Two mobile based platforms are presented, the developed and the Layar application.
Layar can be a useful application for information dissemination. However, it has its
limitations in presenting multiple and time series data sets. Hence, it is needed to
develop a customised application. The front end of the developed mobile application is
designed not only for information access but also to enable the mobile phone users to

send feedback of the lake's actual water quality status. Feedback can be in a form of text, data values or images from the mobile phones camera. The measured data and users' feedback can then be presented both in the main web platform and mobile platform.

Modelling support

The model output was used in the first version of the mobile and web applications for water quality monitoring services. These applications contain monitored data for 61 lakes in the Brabant Delta area, whereas for the Binnenschelde lake, the monitored data are delivered together with the modelled data. The final version of application did not include the model because of its uncertainty and accuracy. In the future, it is envisaged that such water quality model can be integrated in the applications once the issue of uncertainty and accuracy is resolved. The following describes briefly the model and the model-setup.

The water quality model for this study used the open source MOHID modelling package, downloadable at www.mohid.com. MOHID Water Modelling System is a fully nonlinear three-dimensional baroclinic modular finite volumes (finite volume method) water modelling system, created and developed by Instituto Superior Technico (Viegas et al., 2009). It is an integrated modelling tool able to simulate physical and biogeochemical processes.

MOHID modelling consists of usage of different data modules such as atmospheric properties (wind, solar radiation, temperature, precipitation) and bathymetry. There was no established bathymetry for Binnenschelde, bathymetry was created from the average depth data of the lake on a 90mx90m grid.

Water quality modelling for the Binnenschelde lake was set-up to determine the extent of blue algae concentration from the inlet in Zoommeer to the bathing area. This provides supports for decision makers whether to stop pumping from the inlet during the summer season. This decision is important since closing the inlet during the summer lowers down the water level to an extent that bathing is impossible. Combination of different pumping periods and blue algae concentration were modelled. These results can also be used for testing the performance of possible management options in the lake.

Final design

The web platform was designed and customised to fit the users' needs and requirements. After several consultations with representative users, several modifications were made to the first version of web platform before coming up with the final design (Figure 6.2.). The first version of the platform presented the model results, however in the final design it was excluded because of the uncertainty of the model. The final design presents one interface (with the use pop-up windows) for accessing information or sending feedback.

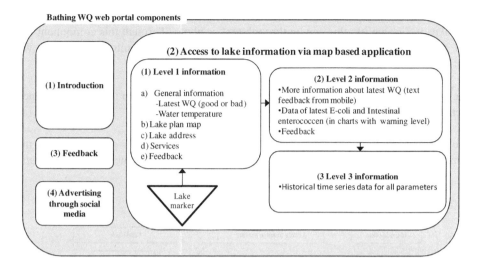

Figure 6.2. Final design: Bathing water quality web-platform for Noord-Brabant lakes
Almoradie (2013)

The web platforms components are the (1) *Introduction*, (2) *Access to lake information*, (3) *Feedback* and (4) *Advertising through social media*. Introduction provides a brief overview of the purpose and use of the platform.

The component *Access to lake information* made use of map based application as the main interface in accessing lake water quality information. Feedback through mobile-phones is integrated in this component. A lake can be located using the search functionality or by browsing the map. When clicking on a particular lake, a tabulated pop-up shows where users can access information in three levels. Since the main users of this platform are the bathers and concerned stakeholders (non-experts), the first level information presents only relevant information for non-experts. Such information is an up-to-date information on health risk warnings or closures of the lake, water temperature, EU classification of the quality of the lake, and more general touristic information, such us the location of bathing areas for small children, restaurant and other facilities. Lakes address and route directions are provided. Clicking the route directions brings users to Google maps driving directions or the Dutch public transportation website. The second level of information is intended for bathers/concerned stakeholders and decision makers/specialists. It presents the latest data of the two most important parameters (E-coli and Intestinal enterococcen) for determining the lakes water quality. These data are presented in coloured bar charts (colours represent WQ if good or bad) with description of warning levels. Furthermore, for both first and second level, access to feedback was made available. The third level information is intended for decision makers/specialists, and it presents historical time series data of all parameters.

The *Feedback* component provides functionality for users to send their observations of the lakes water quality; this can be through text or photo. Through this component users can also send comments and suggestion to improve the monitoring and the web-platform.

The last component is intended for advertisement of this platform through social media such as Facebook, Youtube, LinkeIn, Hyves and Twitter.

The mobile applications' final design (shown in Figure 6.3) presents three main components. Similar to the web-platform it has the main components (1) *Introduction*, (2) *Access to information* and (3) *Feedback*.

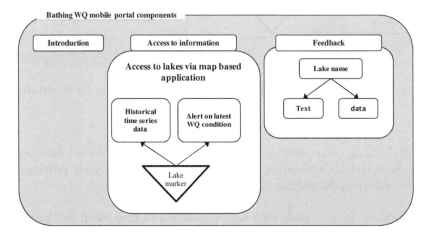

Figure 6.3. Final design: Bathing water quality mobile based platform for Noord-Brabant lakes
Almoradie (2013)

Although quite similar, there are differences in the structure and presentation of information. The structure of presentation and functionality is more oriented for decision makers and experts.   The component *Access to information* presents an alert functionality and access to time series data of all parameters. The *Feedback* component allows users to send both text and data. Text can be a description of the lakes current water quality and data is a measured value (e.g. temperature). Once sent, the data are automatically published in the web platform.

It is important to mention that presented mobile application is also useful for disseminating and gathering information from bathers and concerned stakeholders. However, some modifications are needed in the design and its functionality. These are discussed in the NESP deployment and evaluation chapter.

6.2.2 Implemented design of the Noord-Brabant Water Quality platform

The Noord-Brabant water quality web and mobile platform has been designed to have an efficient front-end and server side application communications. The front-end applications provide an interface for users to access information or send feedback. The server side applications serve all the web content and the needed data. The web and mobile platforms used different technologies in the front end. In the server side both platforms used the same technologies. The components, their communication and the tools and technologies used are presented in Figure 6.4.

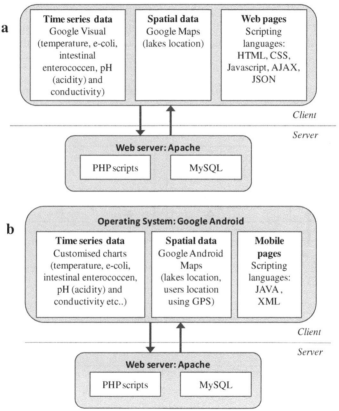

Figure 6.4. Noord-Brabant platform components, communication and used tools and technologies: a) Web platform and b) Mobile based platform
Almoradie (2013)

As shown in the figure both platforms use the same technologies in the server side. The main web server (Apache) provides all the web pages with their basic interactive elements. This web server is extended with several PHP scripts for realizing various tasks on the server side, many of which are related to fetching required data from the implemented mySQL database.

The web platform on the client side has three types of components: Web pages, Spatial data and Time series data. In developing the required interactive user interfaces these components are combined (embedded within each other) on the platform as needed. Server side support is needed for each component in fetching data (e.g. time series data, lake location). Google map was used as the base layer in presenting the lakes spatial location and accessing information. For purposes of visualizing time series data 'Google Visualize' was used for actual graph visualization.

The mobile application, similar to the web platform has also three types of components. It also made use of the Google Maps API. The map applications running on the phone are linked with server-side scripts that provide additional location-specific information, such as time series of measured or modelled water quality parameters. All communication between the client application running on the phone and the server-side scripts are handled using the standard HTTP protocol, which is also used in most web applications.

Figure 6.5 shows clearer presentation of the interactions of these components.

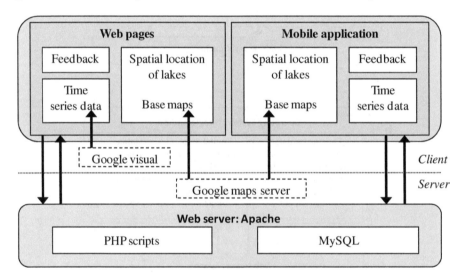

Figure 6.5. Embedding of components: Noord Brabants web and mobile platform
Almoradie (2013)

As shown in the figure the client-server communication takes place in all sections of the web and mobile platform. It shows how base map and spatial locations of lakes are embedded in web pages for accessing water quality information. Visualization of time series data from within such maps is also presented.

6.3   NESP-CP: Danube river and Somes Mare catchment

The following sub-section presents the generic design, modelling support, final design and implementation of the Danube and Somes Mare NESP-CP (web NESP flood platforms). The generic and final design is common for both platforms and serves as a basis for designing the actual web-based user interfaces which are presented in the later sections. However, its implementation had some differences because of additional functionalities used in the Somes Mare.

6.3.1   Generic conceptual and final design

Generic conceptual design

The NESP case application for both the Danube river and Somes Mare catchment is on floods. Based upon the NESP-CP framework the web-based NESP flood platform was designed to be simple, informative, interactive, customizable and a flexible application. Map based applications were extensively used for publishing geospatial data and accessing information. Furthermore, web infrastructure services and data standards were used. Figure 6.6 presents the generic conceptual technical design. The generic conceptual design is part of the Black Sea Catchment Observation System (BSC-OS) which was developed within the enviroGRIDS project. The BSC-OS is composed of several application platforms intended for information dissemination, decision and modelling support. Some platform application made use of Grid infrastructure to run multiple model simulations. The NESP-CP platform did not made use of the Grid infrastructure.

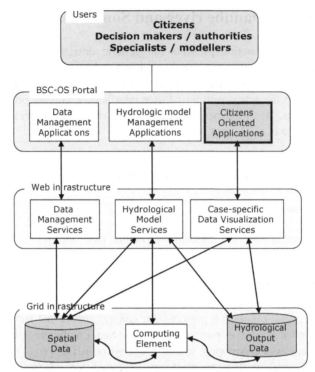

Figure 6.6. Generic conceptual design-NESP-CP: Web-based flood platform for
citizens for the Danube and Somes Mare case study
Jonoski et al. (2013) (Adopted)

It was initially envisaged to make use of the Grid infrastructure to develop a flood forecasting and warning system in the Somes Mare catchment case study. However, it was later found out that it was difficult to calibrate a HEC-HMS hydrological model with snowmelt. The model has difficulty to capture high discharges during spring. If the Grid infrastructure was used it would have run different precipitation inputs to create forecast ensembles of discharge. Worth mentioning is that there was a successful test run of the HEC-HMS model via the Grid, a thousand model instance was distributed and run over the Grid infrastructure. For the Danube NESP flood platform the Grid infrastructure was also not used.

The Web infrastructure was extensively used by both NESP platform to publish spatial data. Additional components that were not envisaged in the original conceptual design were added in both NESP flood platforms. It was realised that for publishing time series data over the web it is beneficial to make use of standards. Using standards provides interoperability of use for modellers and decision makers. The adopted standard by OGC for sharing water-related time series data via web services is the WaterML 2.0 format. Figure 6.7 presents the updated general conceptual design.

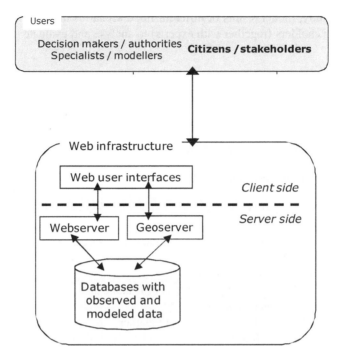

Figure 6.7. Updated generic NESP flood platform conceptual design
Jonoski et al. (2013) (Adopted)

In Figure 6.7 the NESP flood platform web infrastructure is composed of a web based user interface in the client side and three main components in the server side: (a) Databases for observed and modeled data, (b) Geoserver to archive and publish geospatial and times series data in standard format and (c) Web server.

Modelling support

The following presents the modelling support for the Danube and Somes Mare case study.

a)   Danube river (Braila-Isaccea section)

The SOBEK 1D/2D was used as the modelling support for the NESP. SOBEK is a licensed modelling software developed by Deltares. The software tool is used for many areas in flow modelling such as irrigation systems, drainage systems and natural streams. The SOBEK 1D/2D couples one-dimensional (1D) and two dimensional model (2D); 1D represents the river channel and the 2D represents the floodplains. All data necessary for the Danube NESP platform and model have been provided by National Institute for Hydrology and Water Management of Romania (INHGA).

In this case study, model results of different flood scenarios have been deployed in the NESP for stakeholders (together with experts) to analyse and evaluate the pressures and impacts.

The 1D/2D SOBEK model has been set-up to simulate flood event scenarios in the Braila-Isaccea. The main aim is to simulate different flood peaks (from Danube, Siret and Prut) and dike scenarios (without and with rising of the dikes). The model developed is envisaged to be further developed in the future as a tool for learning about the pressures, impacts and mitigation strategies that can be implemented in the area.

The modelling for this case study has been carried as follows:

1. A 1D hydrodynamic model was set-up and calibrated to simulate the channel river flow.
2. Coupling of the 1D and 2D model to simulate flooding in the floodplain.
3. Simulating flood scenarios:
    - Scenario 1: Danube (base case)
    - Scenario 2: Coincident peak arrival of Danube and Siret
    - Scenario 3: Danube and Prut
    - Scenario 4: Danube, Prut and Siret together
4. Simulating all above flood scenarios with the rising of the dikes by 3 m.

A flood event corresponding to about 1/100 years return period was selected for simulation. This flood event is a period from 18 June 2010 to 10 July 2010. The Prut and Siret measured flood hydrographs have been shifted (coincidental arrival of flood peaks) to simulate the simultaneous flooding scenario with the Danube river. A measured hydrograph (Q) in a station near Braila was used as the upstream boundary condition. A rating curve (Q-h) at Isaccea was used as the downstream boundary condition. The 1D model was calibrated by adjusting the roughness coefficient in the river bed. After calibrating the 1D model, 1D-2D simulations have been carried out.

A sample snapshot of flood maps are presented in Figure 6.8, presenting scenarios of coincidental flooding from the Danube, Siret and Prut where dikes are not raised (A) and dikes raised by 3m (B). These results are taken from the actual NESP platform.

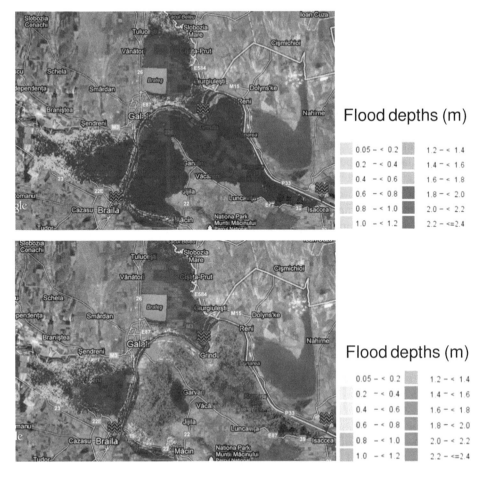

Figure 6.8. Flood maps for scenarios A (top) and B (bottom)
Jonoski et al. (2013) (Adopted)

b)   Somes Mare catchment

   In this case study a HEC-HMS rainfall-runoff model was set-up as a supporting tool
for the NESP. HEC-HMS is a hydrological modelling systems developed by the US
Army Corps of Engineers. Most of the data used for rainfall-runoff modelling were
provided by the Romanian Waters - Directorate of Somes-Tisza (DST). The Romanian
Waters – Directorate of Somes-Tisza are responsible for water and flood management in
the area.

The Somes Mare catchment was divided in 26 sub-basins. A discharge varying in time
was introduced as an input coming from the reservoir at Bistrita. The model made use of
the 26 rainfall stations, 23 temperature stations and 19 discharge stations in the
catchment; 3 of the discharge stations are along the main river. Gage weights provided
by expert hydrologists from the Somes Tisza Water Branch of Romanian Waters (the

flood management authority) were used to distribute precipitation over the sub-basins. Monthly evaporation was used in the model. Simple canopy, simple surface, Soil Moisture Accounting (SMA) loss method, SCS unit hydrograph for the transform method and linear reservoir for the base flow method were used for all sub-basin. The lag routing method was used for routing along the river reaches. Since snowmelt highly influences the discharge on Somes Mare, the Temperature Index snowmelt method was introduced in the model.

The model has been calibrated for the period 2006-2008 since air temperature data are only available from period 2006-2008. Air temperature is significant for the snow melt component of the model. Shown in Figure 6.9, the model performed satisfactorily on the runoff pattern however it performed very poor in simulating the flood peaks. Further adjustment of its parameter specially related to snowmelt is needed for it to be used in the future for flood forecasting.

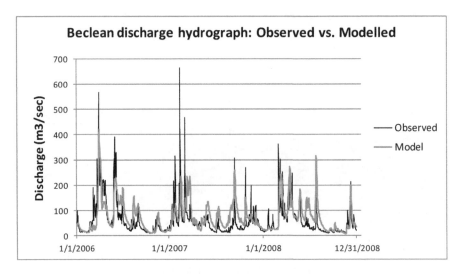

Figure 6.9. Simulated and observed hydrographs for Beclean station (2006-2008)
Almoradie (2013)

Even though the model results require further improvement it was decided to include them in the NESP platform, as they may still provide useful information for both water professionals interested in this kind of modelling as well as for other citizens and stakeholders.

Final design

The final design of the NESP-CP flood platforms was designed to have several sections that are straightforward for Citizens to use. The first sections contain the introductory pages explaining the purpose of the platform, usage and information about the study area. Another section also provides information on flooding problems and existing flood management strategies. The most interactive section is the actual flood platform data and modelling results. Citizen's feedback on flood-related issues is one of the components in the NESP flood platform section. Figure 6.10 is the final design of the front end of the platforms

Flood portal components

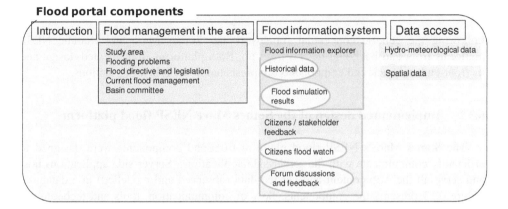

Figure 6.10. Final design of the front-end of the NESP flood platforms presenting its components
Jonoski et al. (2013) (Adopted)

The section "*Introduction*" provides users information on the purpose of the NESP flood platform, why it was developed and how to use it. The section "*Flood management in the area*" presents information of the study area, its flooding problems, flood directives and legislation (EU FD), the existing flood risk management plans and strategies and the basin committees. The Basin committees represent the interest of the people and its role is to communicate with different stakeholders. Further, in this section user will be able to know the major stakeholders involved in flood related issues.

The "*Flood information system*" is the main section of the platform where users can access historical floods, model results and send feedback. The section is highly interactive and rich in information. The section consists of two main components: (1) "*Flood information explorer*" and (2) "*Citizens / stakeholders feedback*". The "Flood information explorer" has two sub-components "Historical floods" and "Flood simulation results". "Historical floods" presents a map-based interface to access and view historical time series data in charts. Flood thresholds and an example of a flood event are presented to provide better understanding for users. The "Flood simulation results" has similar method in presenting modelling results from the developed models (different for the two different case studies). The other main component "Citizens / stakeholders feedback" provides different possible means to send feedback. An important functionality in this component is the use of a map based interface to send flood-related information, with this user can send messages or pictures of the flooded area. Discussion forum, and forms are provided for users to send feedback about the platform. For the Somes Mare platform additional section has been added to present the set-up and explanation of the HEC-HMS model, this was intended more for professional users and not the citizens.

The last section *"Data access"* is also intended for professional users such as decision makers and modellers. The actual hydro-meteorological, time series and spatial data can be access through this section. OGC standards such as WMS, WFS and WaterML were used in publishing these data.

Following the generic conceptual framework the Somes Mare and Danube NESP platforms front and back-end has been set-up. Both platforms implemented design and its front-end and back-end components are presented in the following sections.

6.3.2   Implemented design of the Somes Mare NESP flood platform

The Somes Mare's NESP flood platform front-end components were designed to efficiently communicate with the server side applications. Server side applications host and serve all the web content and needed data (observed and modelled) in a database. Figure 6.11 presents the components, flow of communication, tools and technologies for the NESP flood platform.

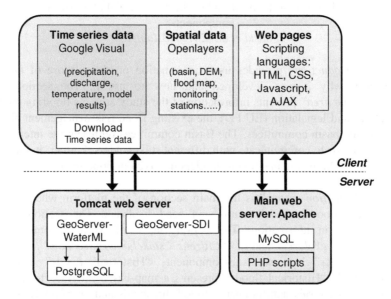

Figure 6.11. Somes Mare platform components, communication and used tools and technologies
Jonoski et al. (2013) (Adopted)

The client side has three main components, the Web pages, Spatial data and Time series data. These components are combined as needed to achieve the required interactive interfaces. Server side components provide supports for the three front-end components. The server side consist of the main web server (Apache) and another server (Tomcat). The main server hosts all the web pages. It makes use of PHP as the scripting language and MySQL as the database system. The Tomcat server was set-up for the tools for publishing spatial and temporal data (GeoServer).

Map interfaces were embedded for many pages especially in the Flood Information System. Map interfaces were extensively used to present spatial information such as the study area's DEM, spatial flood model results and monitoring stations. Google maps were used as the base maps and spatial data are published by Geoserver application named "Geoserver-SDI". Openlayers (javascript libraries for visualising and manipulating spatial data) was used to have an interactive web map application. Accessing time series data is done through a request from the client side via the map based interface. The client sends a request to the server then to the database to extract and send back the requested data. This is then received and processed by the client and displayed using the "Google Visualize" graphing component.

The last component "Download time-series data" presents also a map based front-end interface to select monitoring stations and download time series data in WaterML 2.0 format. The framework and methods by the Commonwealth Scientific and Industrial Research Organization (CSIRO) in Australia that implements the WaterML 2.0 schema using the Geoserver Web Feature Services (WFS) was applied in this platform. In summary it requires an instance of specially configured Geoserver ('Geoserver-WaterML' in Figure 6.12) and setting up a PostgreSQL database.

Figure 6.12 shows typical embedding cases presenting more detailed interactions of these components.

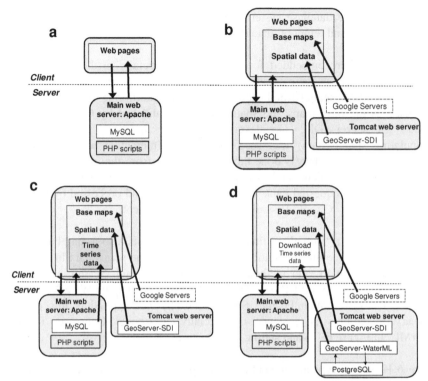

Figure 6.12. Embedding of components (Somes Mare): a) standard web pages; b) spatial data embedded in web pages; c)Time series data visualized from maps in web pages; d) Downloading time series data
Jonoski et al. (2013) (Adopted)

Case (a) has the simplest form of communication which takes place in all sections of the platform. Case (b) presents embedding of maps in web pages and accessing of spatial data via the base map. Case (c) shows access of time series data with in maps. These two types of communication are in most of the pages of the "Flood Information System". Lastly, Case (d) presents how communication takes places in the "Data access" section, downloading time series data in WaterML 2.0. As previously mentioned an instance of Geoserver was set-up with modifications to publish time series in WaterML 2.0 format.

6.3.3   Implemented design of the Danube NESP flood platform

The implemented design of the Danube NESP flood platform is similar to the Somes Mare platform. The Danube platforms difference is the implementation of time series data download. Here Geoserver-WaterML was not implemented. Figure 6.13 presents the components and their communication.

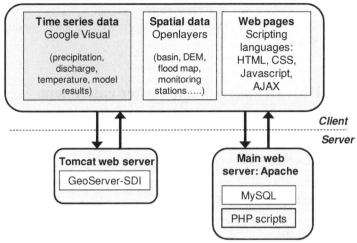

Figure 6.13. Danube platform components, communication and used tools and technologies
Jonoski et al. (2013) (Adopted)

Figure 6.14 presents only the three cases of embedding of components.

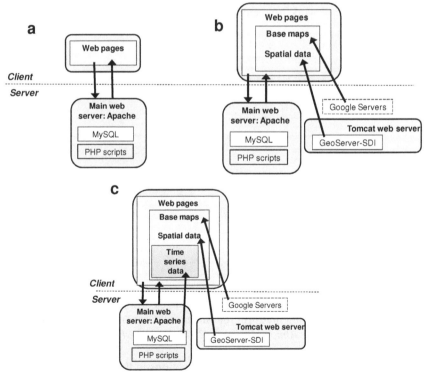

Figure 6.14. Embedding of components (Danube): a) standard web pages; b) spatial data embedded in web pages; c)Time series data visualized from maps in web pages
Jonoski et al. (2013) (Adopted)

For the rest, the description of the components, their embedding and communication is same as for Somes Mare.

6.4   NESP-CDM: Cranbrook and Alster catchment

The following sub-section presents the generic design, modelling support, final design and implementation and decision methods used of the NESP-CDM for both Cranbrook and Alster case studies.

6.4.1   Generic conceptual and final design

Generic conceptual design

The NESP case application for both the Cranbrook and Alster catchment is on flood risk management. The collaborative platform was designed based upon the NESP-CDM framework introduced. In addition to all the necessary functions the platform was designed to be easy to follow, interactive, customisable and flexible enough to be implemented in other case studies. Figure 6.15 presents the generic conceptual design that shows the general workflow and the client-server communication of the envisaged components.

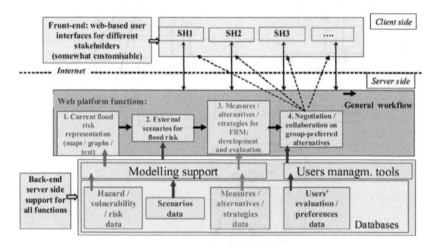

Figure 6.15. Generic conceptual design- NESP-CDM: Collaborative modelling and decision making platform for Alster and Cranbrook case study
Almoradie et al. (2013) (Adopted)

This figure also presents the interactions and functions in the front-end and back-end of the NESP-CDM. Through the front-end interface, individual stakeholders can access the web-platform to learn about the study area and current flood risk, external scenarios, develop and evaluate measures and alternatives, and, lastly, they can negotiate towards a consensus on proposed alternatives to be implemented. These are all supported in the back-end by models and management tools.

Modelling support

The following presents the model and model set-up for the Cranbrook and Alster catchment.

a)   Cranbrook catchment

In urban areas, rain falls initially on surface areas, flows towards the manhole or drainage then enters the sewer system. During an extreme rainfall event the maximum storage capacity of the sewer system may be filled to the point that water overflows to the surface causing pluvial (surface flooding). Such pluvial flooding occurs in the Cranbrook catchment.

The Cranbrook case study combined the physically-based and dual drainage surface models to simulate pluvial flooding in the catchment. The flood models were set up and calibrated in Infoworks CS by the project partners (Imperial College of London, UK). These models are: (1) 1Dimensional-2Dimesional (1D-2D) model for mapping purposes and (2) 1D-1D model for optimisation of computational time in future forecasting applications, which includes the AOFD (Automatic Overland Flow Delineation) techniques (Maksimović et al. 2009). The second model is envisaged to be used in future as a forecasting tool (combined with rainfall forecasts) considering that urban areas get flooded in 10 to 20 minutes by pluvial flooding when heavy rainfall occurs. A methodology proposed by Wang et al. (2011) in cooperation with the MetOffice (UK weather forecast) was introduced to further enhance the model forecasting capability for flood event management.

In urban pluvial flood modelling, to have a more accurate result it is necessary to make use of data that realistically represent the urban landscape. When urban flooding occurs there is a need to take into account the dynamic interaction between the overland system and sewer system. The concept of interaction between the overland and sewer system is known as the "dual drainage concept". The developed model adopts such concept, which consists the "surface network model", "sewer network model" and "dual-drainage model". Figure 6.16 presents the Dual-drainage concept.

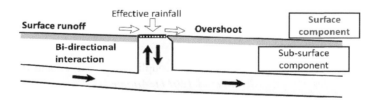

Figure 6.16. Dual-drainage concept
ICOL (2011)

Some of the model results are presented in Chapter 7, which is about the deployment of NESP.

b)   Alster catchment

Two flood models were used for the Alster case study, the MIKE 11 and HEC-RAS model. MIKE 11 developed by Danish Hydraulic Institute (DHI) is a licensed software package for one dimensional (1D) flow simulation. It includes modules for simulation of water quality, sediment transport in estuaries, rivers, channels and irrigation systems. The hydrodynamic component is the heart of MIKE 11. It can also simulate river flows with structures (e.g. sluice, dams, spillways, culverts and bridges). The HEC-RAS modelling tool is similar to MIKE 11. The main difference is HEC-RAS is a freely available river modelling software. HEC-RAS was developed by the US Army Corps of Engineers (USACE).

The project partners made use of an existing MIKE 11 model (provided by LSBG) for the upper part of the catchment and coupling it with a HEC-RAS model for the lower part. The HEC-RAS model was set up by the project partners (Leuphana University of Lunenburg, German).   The upper part of the catchment was modelled until the Fuhsbüttlerschleuse (Figure 6.17), while the lower part was modelled from Fuhsbüttlerschleuse to Schaartoschleuse. Figure 6.17 presents the coupling of Mike 11 and HEC-RAS.

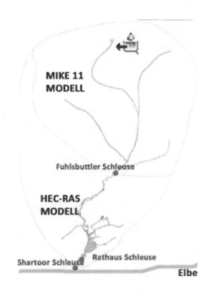

Figure 6.17. Alster catchment models: Coupling of MIKE 11 and HEC-RAS
Evers et al. (2011) (Adopted)

LSBG provided the required input data. The required input data for the HEC-RAS model are the river length and cross-section, structures along the river, discharge and water level measurements, gates and pumping stations operation data. The MIKE 11 output was used as the upstream boundary condition and the tidal curve from the Elbe was used for the downstream boundary condition.

FRM guidelines for the city of Hamburg (LSBG, 2009) have been followed for the preliminary flood risk assessment. Return periods of 100, 200 and 300 year return period were simulated. In the end the 200 return period was used as a base case. Some extreme scenarios in the city of Hamburg associated with failure of tidal protection structures in combination with 200 year return period were also modelled, but they were not used in NESP platform.

The NESP platform made use of the model results and the so called identified "hot spots" for stakeholder participation.

Final design

The final design of the platform resulted in two modules named "*Collaborative Platform*" (CP) and "*Collaborative Modelling Exercise*" (CME), corresponding to the two stages of the collaborative framework (Figure 6.18). The CP was designed for most of the collaborative modelling activities and the CME for the participatory decision making. Components for visualisation and interactivity were made available through maps, tables, graphical charts, forums, chat, feedback forms, pre-prepared videos and flowcharts. Figure 6.18 gives a summary of the CP and CME design as implemented and used in the case studies.

The Collaborative platform consists of (1) *Introduction*, (2) *Study area* - information about the study area and description of the models used, (3) *Flood risk framework* - information about key terms related to flood risk, (4) *Flood risk management* - framework for managing flood risk, (5) *Stakeholder* - information about the stakeholders and (6) *Collaborative modelling* - forums, feedback and link to the *Collaborative modelling exercise*.

The Collaborative modelling exercise consists of (1) *Introduction* – users' guide on how to use the platform, (2) *Individual profile* - a component where stakeholders evaluate the proposed alternatives with respect to identified objectives so that they can obtain individual ranking of the alternatives and (3) *Group profile* - a component where individual rankings are aggregated for the whole group and presented in such a way that individual positions of a stakeholder within the group are made as transparent as possible. A chat component for instant communication among the stakeholders is also provided.

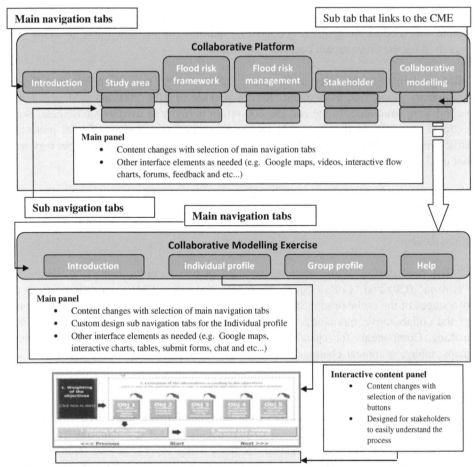

Figure 6.18. Front-end final design of the CP and CME as implemented in the case studies
Almoradie et al. (2013) (Adopted)

6.4.2   Implemented design

Implementation design for both Cranbrook and Alster case study are similar. The platforms were designed to have an effective and efficient communication of all components in the front-end with the server side applications.   The implemented components and their communication are presented in Figure 6.19.

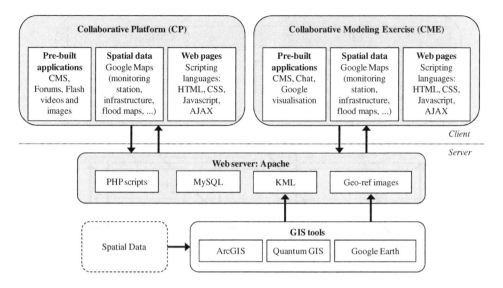

Figure 6.19. Alster and Cranbrook platforms components, communication and used
tools and technologies
Almoradie (2013)

There are three types of components in both the CP and CME platforms client side: web pages, pre-built applications and spatial data. These components are combined as needed to create an interactive interface in the web pages. Such interactive pages are the presentation of spatial information with the use of Google map. Javascript was used in implementing the TOPSIS method on the client side as part of the judgement engine. The judgement engine was designed to be deployed in the client side in order to acquire the results faster without using server side applications, which were invoked only for storing the submitted individual profiles.

The client side components need sever side support to host and store databases. The server side used the Apache server with the same technologies (PhP scripts and MySQL) like the other case studies presented in this chapter. The only difference is in the storage and format of spatial data. Unlike the case studies presented in Section 6.3, the Alster and Cranbrook case study demonstrated that it is in fact possible to present raster spatial data without the use of standards, like those proposed by OGC for building SDI. The spatial data format hosted in the server side in this case is the Keyhole mark-up language (KML) and Geo-referenced image. Unlike the SDI which is interoperable, this approach requires pre-processing, which in this case was done by using several GIS tools.

Figure 6.20 shows clearer presentation of the interactions of these components.

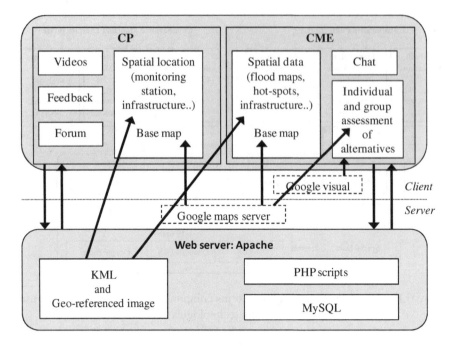

Figure 6.20. Embedding of components for the Alster and Cranbrook catchment
Almoradie (2013)

The figure shows how all web pages are provided in the platform. This communication takes place in all sections of the CP and CME. Also shown in the figure is how maps are embedded in web pages for accessing spatial data on top of base map layers. In the CME, visualisation using charts were extensively used, and Google maps were customised for mapping the positions of individual's rankings of alternatives in the Group profile. This type of visualisation was named "Swimming pool of alternatives" and will be demonstrated in the following chapter.

6.5   Concluding remarks

The design and implementation of NESPs were to some extent similar for all case studies. The differences come primarily from the case study characteristics (type of problem, level of participation), and to some extent from the requirements of the research projects within which this work has been carried out. (For example, the demonstration of the potential of SDI technologies and WaterML was a requirement within the enviroGRIDS project).

In all cases the design and functionalities of applications were developed through consultations with the stakeholders (and technical partners) and to some extent the citizens (through organised workshops). With this approach the developed applications acquire higher potential to be used in practice. This approach is recommended by Evers

(2008) as one of the requirements when developing DSSs in integrated river basin management.

The selection of available GPL technologies must be done carefully following a set of criteria (presented in Chapter 5.). Experiences in the development of the platform with the use of GPL technologies are important for gaining insights about their flexibility for adoption in other case studies or field of application. Moreover, the developers' experiences may provide technological foresight, a kind of exploration of how new and future information technologies will further shape stakeholder participation in flood risk management.

The actual experiences in the work presented here can be stated that the Apache server and MySQL database were very easy to install and configure and crashes were never encountered during the development and working with the platform. These two technologies are widely used by the international community. Two of the giants in social networking use such technologies namely Tweeter and Facebook.

Standard scripting technologies such as the HTML, CSS, Javascript and AJAX were very flexible in structuring and implementing the web interfaces and customising the navigation and interactivity of contents. Especially Javascript technology performed very well for the Judgement engine (TOPSIS method). The PHP server side scripting language is efficient in querying the database, computing and sending back the required information.

The pre-built components such as the CMS, forum, chat and Google maps were flexible and compatible to be combined in the platform. Since these pre-built components were connected using the same standard scripting technologies there were no problems in integrating them with the server and the database. However, freely available Google maps API has some limitations, such as the number of geocode requests per day (2,500 geolocation/per day) and the size and complexity of loaded KML files (max 10MB). In the development of the platform, there were polygon features in KML file format that were more than 10 MB and because of these limitations the polygon features were divided in several KML files (each less than 10MB) or processed in geo-referenced images. Dividing the polygon feature into several KML files may be too tedious if there are too many polygons, such as the results of the flooding models in the Alster and Cranbrook case studies, thus it was opted to use high-resolution images and geo-reference them in Google maps. Pre-processing was done using ArcGIS, Quantum GIS, imaging tools and Google earth.

The implementation of web services (WMS, WFS, WFS-WaterML 2.0) using GeoServer technologies combined with web interfaces has been smoothly carried out. However, the GeoServer-WaterML 2.0 implementation was not that straightforward and it required several back-end configurations for achieving the required functionalities. Furthermore, there was a need to customise the GeoServer's front-end for WaterML 2.0.

In the development of the NESP applications, programming skills are not sufficient by themselves. Interdisciplinary knowledge and skills are required, usually available in teams of developers, which fortunately was available, as all research projects were collaborative in nature with different research partners.

For future applications the potential of new technologies that are currently in their infancy or early stages (e.g. HTML 5) may need to be considered. The explosive growth in usage of mobile smartphones may also need to be taken into account for developing future applications for stakeholder collaboration in flood risk management.

In general the use of GPL technologies for such platforms is highly feasible. They do provide the desired level of interactivity in the developed components and have the flexibility to be adopted in other case studies.

Chapter 7
Deployment and Evaluation of NESPs

*"The ultimate value of life depends upon awareness and the power of contemplation
rather than upon mere survival."*

~Aristotle

This chapter presents the deployment and evaluation of NESP. It starts with a brief overview of evaluation methods and their application in this research. Afterwards the deployment and evaluation of the NESPs for the five case studies are presented. Lastly are the summarised conclusions for the different NESPs.

7.1 Deployment methods

In general, the deployment of all NESPs was organised with a face to face workshop with the stakeholders. For some case studies it was followed by a series of workshops.

The first workshop with the stakeholders was intended to introduce the platform and its objectives and to demonstrate its use. The introduction and demonstration were then followed by a hands-on use of the platform by the stakeholders. Here the stakeholders explore and familiarize themselves with the platform functionalities. After the workshops stakeholders were encouraged to make use of the platform for approximately a month or more.

Evaluation of the platforms was carried out after the workshop. Evaluation methods used are presented in the following section.

7.2 Evaluation methods

Evaluation of the NESP and its participatory process is essential to gain insight on the strengths, weaknesses and value of the networked environment implemented, as well as on the technologies used. This can help improve the NESPs in their future implementation. Moreover, through this evaluation, conclusions can be drawn out as to whether the networked environments can address the hindrances and potential pitfalls in stakeholder participation.

Several evaluation methods have been developed for stakeholder participation, such as those by Rowe and Frewer (2004), Rasche et al. (2006) and the COPIR (Constraints, Objectives, Process, Intensities, Reporting) approach by Krywkow (2009).

Rasche et al. (2006) introduce the use of intensity criteria for evaluation. The evaluation approach of Rowe and Frewer (2004) aims to identify the most appropriate methods of a participatory process while the COPIR approach goes a step further by introducing altogether the goals of the participatory process and the planning goals. This approach adopted some of the methods of Rasche et al. (2006) and additional methods have been added.

In this study, the evaluation of successful stakeholder participation is case specific and it is difficult to have a generic evaluation method. The definitive evaluation is when the stakeholder participation for water resources and flood management is implemented in practice. However for research purposes this needs to be carried out through carefully designed workshops and exercises of stakeholder involvement, when the developed NESP are tested. The actual evaluation is carried out by the involved stakeholders.

The intensity criteria evaluation method of Rasche et al. (2006) was used for the Noord-Brabant case study. For the other case studies evaluation results were presented in bar charts.

In the end of the workshops participants received evaluation forms with a set of evaluation questions that they were requested to answer. The evaluation forms contain questions that are answerable in quantifiable scales (values or linguistic descriptions) or in specific answers (e.g. which visualisation tools you find useful).

The subsequent sections presents the NESP deployment, followed by the results from the stakeholders' evaluation.

7.3   NESP-IKS: Noord Brabant lakes

The demonstration and testing of the water quality web-mobile based applications were carried out in two stages. The first stage was the continuous user (mainly professional users) involvement and feedback combined with interactive development of the applications. The second stage was focussed more on public user testing. In the summer of 2011 face-to-face workshops were held with the public users, during which structured evaluation questionnaires were handed out.

The subsequent sections present the deployment and the summarised results of stakeholders' evaluation.

7.3.1   Deployment

The presentation in this section is structured in two parts: Deployment of (1) web-based application and (2) mobile application.

Web-based application

Intended for public users, a website was developed for the Brabant Delta to communicate bathing water quality of lakes. The design of the website was simple, straightforward and clear for users to easily understand the purpose of the website and to navigate the maps for lakes in the Brabant Delta. Sixty one lakes in the area are presented in this platform. Figure 7.1 presents the main front-end of the web platform.

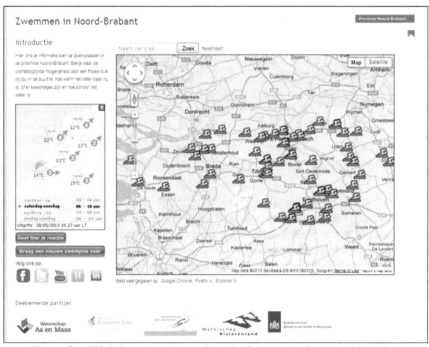

Figure 7.1. Web-based water quality platform main front-end interface
Almoradie (2013)

Access to the platform is provided in the following web link (password protected):

* http://hikm.ihe.nl/Lenvis/Noord_Brabant

Clicking on a particular lake shows the first level of information (intended for citizens). A pop-up box with tab menus is made available presenting up-to-date information of water quality warning, water temperature, EU classification of the quality of the lake. Also presented is general touristic information, such us the location of bathing areas for small children, restaurants, water activities and first aid and safety facilities. Lakes address and route directions linking Google map driving directions and Dutch public transport sites are also provided (Figure 7.2).

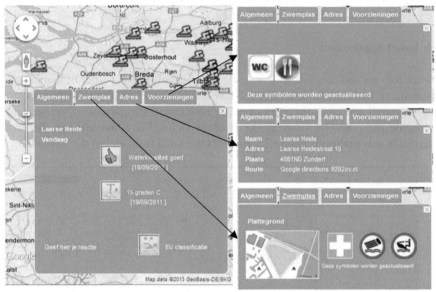

Figure 7.2. Noord Brabant water quality web platform: first level information (for citizens)
Almoradie (2013)

Clicking on the water quality alert (hand icon or the text) gives users access to the second level information (for citizens and professionals) (see Figure 7.3). A pop-up window presents a bar chart of the latest measured value of the two most important indicators for bathing water quality (*Intestinal enterococcen and E-coli*). The bar chart changes colour if the parameter is normal (blue) or above limit (red). Also presented here are user feedbacks in text format (for general users) sent from the field via mobile phone.

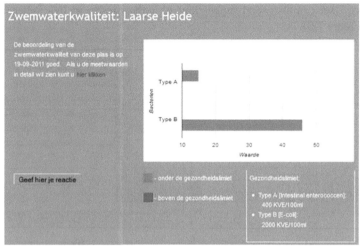

Figure 7.3. Noord Brabant water quality web platform: second level information (for
citizens and professionals)
Almoradie (2013)

A link provided in the second level information gives access to the third level
information (intended for professionals). In a pop-up window the user can access
historical time series data of intestinal enterococcen, e-coli, water temperature, acidity
and water conductivity. The historical time series chart also presents data values (for
authorized users) sent from the field via mobile phone (Figure 7.4).

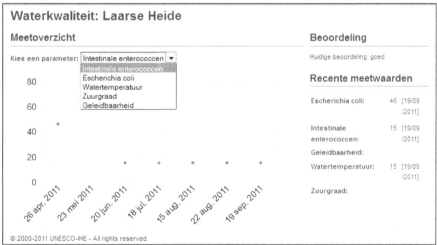

Figure 7.4. Noord Brabant water quality web platform: third level information (for
professionals)
Almoradie (2013)

Web based user feedbacks are also provided in the platform. Users can give general
reaction about the website or comments about a specific lake (they can send both text

and photos). User can also send suggestions to authorities to include other lakes in the platform (Figure 7.5).

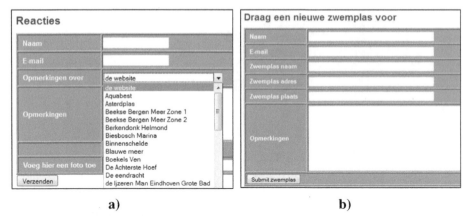

a)   b)

Figure 7.5. Water quality web-based feedback: (a) comments for the website or a lake
(b) suggest new lake
Almoradie (2013)

Mobile application

Similar to the web platform the mobile application deals with dissemination of measured water quality information of 61 lakes which are being used by the general population for swimming, bathing, surfing, etc. Besides access to water quality information, the mobile application can also send data collected by the authorities and feedback on actual water quality status which can be either in a form of text or images via phones' camera. These information (measured and users' feedback) can then be presented in on the website. Information that is made available in the website can also be viewed via the mobile phone itself. Android OS mobile device were used to develop the application.

The application has three main sections accessible via tabs: (1) *introduction*, (2) *access to lakes water quality status and historical data* (3) *feedback*. Figure 7.6 presents a snapshot of the first tab of the application.

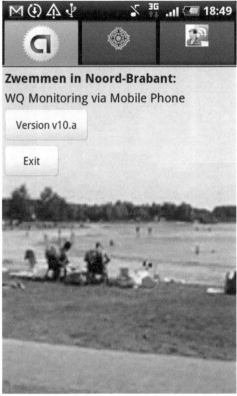

Figure 7.6. Noord Brabant Mobile application: Introduction
Almoradie (2013)

In the second tab a map-based interface with the lakes location is presented. Once a lake is selected, users can access the lakes water quality status or its historical time series data (Figure 7.7).

Figure 7.7. Noord Brabant Mobile application: Access to lakes water quality status and
historical time series data
Almoradie (2013)

The third tab presents an interface for provision of user feedback. Feedback can be
either textual information or data values. As previously mentioned this information
provided by the users/citizens (textual information) and professionals (data) appears on
the dedicated website together with all other relevant water quality information for the
area provided by the authorities (Figure 7.8).

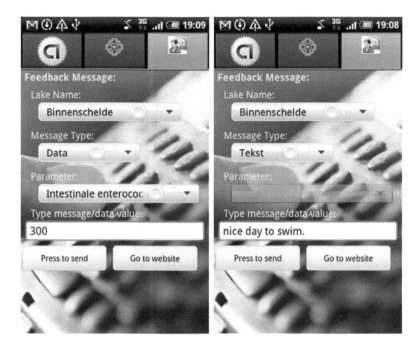

Figure 7.8. Noord Brabant Mobile application: Feedback- data and text
Almoradie (2013)

One of the most interesting separate functionality in the application is the use of augmented reality browser named "Layar". The lakes and its accompanying information (name, water quality status and address) are presented in the landscape via mobile phones camera. Layar also makes use of GPS and digital compass to give directions and pin-point locations of point source features. All relevant information such as the lake coordinates, name and water quality status are stored and accessed in a database.

The application can be accessed through any mobile devices with Layar application using the search word "Zwemmen N. Brabant" A screenshot of the Layar interface for the Brabant Delta application is presented in Figure 7.9.

Figure 7.9. Noord Brabant Mobile-Layar application: (from left to right) search application using keywords "Zwemmen N. Brabant", the augmented reality browser showing the lakes, list of lakes in table form, lakes in Google maps, and lastly driving directions.
Almoradie (2013)

Layar is a promising application to disseminate water and environmental-related information. It is more beneficial if future applications combine both observed and model results in the natural landscape via the mobile phones camera.

7.3.2 Stakeholders evaluation

Evaluation process

As mentioned, the applications presented have been evaluated and validated by both public and professional users. During the developmental stage continuous professional users' involvement and feedback were included in the design and functionality of the applications. The provinces, waterboards, national health, environment and water organisations, and consultancy bureaus were greatly involved in the developmental stage.

The first version of the developed application was presented and tested by the users during the face to face meetings. Feedbacks were taken such as its ease of use, functionality, suggested changes or additions. Additional feedbacks were also communicated via e-mail. Main feedback was received during the mid-term seminar of the LENVIS project were a large and diverse user groups made a hands-on test.

The next step of evaluation proceeded with the public users test. Face to face meetings with selected public users made a hands-on test. Feedbacks were provided during the hands-on test and at the end structured evaluation questionnaires were given to the users.

The intensity criteria evaluation method of Rasche et al. (2006) was used for the integrated web-based computer and mobile application.

Evaluation results

Five professional users and eight public users tested and evaluated the applications.

Overall the users gave a positive feedback or evaluation for both the web-based computer and mobile applications. One of the tests that have a positive feedback is the user friendliness of the applications. Based from users experience they found the applications simple to navigate and are easy to understand.

Mobile application:

Regarding the different types of mobile phone feedback, it shows that structured and unstructured mobile phone feedback applications were considered very useful. However, it is suggested that an alert function should be added if the feedback has been sent successfully or not, or some kind of a progress indicator.

Users also provided feedback on some areas that can still be improved, such as the poor readability of information provided in the mobile phone application. Dedicated mobile phone functionality (GPS, location based services etc..) should be used more else the web application accessed via the mobile phone browser is already sufficient.

Web-based computer application:

During the meetings one of the main conclusions of both the public and professional users was on the type and level of presentation of bathing water quality information. For professional users the use of graphs and numerical data to present water quality information was found to be suitable while for public users qualitative presentation (e.g. 'good', 'bad', 'swimming not advisable', 'swimming not allowed') of water quality was more appropriate.

Another conclusion is on the provision of water quality warnings. Automatic provision of water quality warnings is not advisable since errors in measurement may happen; most often it is difficult to automatically detect these errors in the field. Instead it is recommended that authorized experts (Water boards) should do the interpretation and decision makers (province) should publish the warnings. The province is responsible for issuing warning or temporary closing of the lake.

Intensity criteria- Integrated web-based computer and mobile application:

The evaluation of the water quality website was based on five indicators: system quality, information quality, user impact, user satisfaction, and information use. Validation results of these five indicators, by public and professional users are presented in Figures 7.10 and 7.11, respectively. The scale used for representation is 1–5(1 – the lowest value, 5 – the highest value).

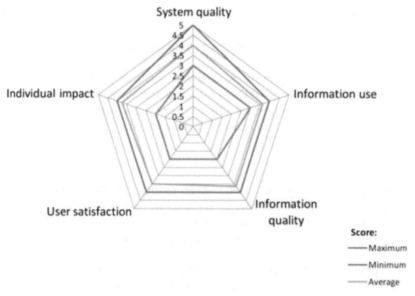

Figure 7.10. Evaluation of the water quality integrated web-system by public users. *Jonoski et al. (2012) (Adopted)*

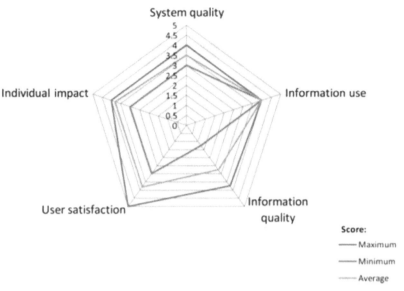

Figure 7.11. Evaluation of the water quality integrated web-system by professional users.
Jonoski et al. (2012) (Adopted)

To summarized, the overall evaluation is positive. Members of all user groups have expressed their clear appreciation for the integrated web-system, and in particular for the mobile applications.

7.4   NESP-CP 1: Somes Mare catchment

Upon finalization of the NESP flood platform for Somes Mare the partners involved in the development organized a workshop with local stakeholders in Romania. The main purpose of the workshops was demonstration of the NESP flood platform to these stakeholders and their evaluation of the platform as a whole and of its different components. The workshop that lasted approximately half a day was organized at the end of February 2013 in Cluj Napoca.

7.4.1   Deployment

The Somes Mare NESP flood platform was named Somes Mare flood platform for citizens. English and Romanian versions of the platform were developed. Following the framework and conceptual design, the Somes Mare FIS was structured in four main sections as follows: (1) *Introduction*, (2) *Somes Mare flood management*, (3) *Flood information system* and (4) *Data access*. Figure 7.12 presents the interface to access the sections and components of the NESP flood platform.

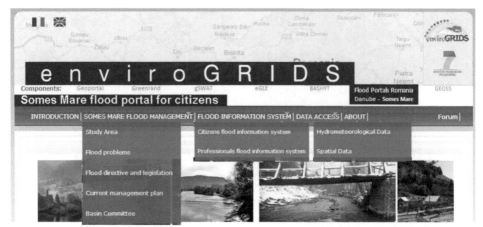

Figure 7.12. Interface to access the sections and components of the NESP flood
platform (Somes Mare case study)
Almoradie (2013)

Full access to the platform is provided in the following web link:

- http://hikm.ihe.nl/envirogrids/platform/somes/

The workshop started with the introduction of the platform and its objectives followed
by a demonstration in using the platform; presenting information, functionalities and
components accessible in each section. Afterwards users were given an opportunity to
use and explore the platform for approximately an hour. During the hands-on several
questions were raised regarding the platforms functionality in accessing the data, such
as using the graphical interface for the presentation of time series. Overall the workshop
went smoothly. The stakeholders and partners generally see the platform as very useful.
The Romanian Waters - Directorate of Somes-Tisza were interested to make use of the
NESP flood platform system as component in their flood management.

The following presents a summary of the results from the NESP flood platform
application.

The *Introduction* section presents information about the platforms purpose and use.
It also provides direct links to the discussion forum, feedback forms and contact
information.

Somes Mare flood management

Raising citizens' awareness starts with the Somes Mare flood management section
(Figure 7.13). Providing information of the study area, its' flooding problems, flood
directives, current FRM plan and the major stakeholders. They are available on the
platform and they will not be presented here.

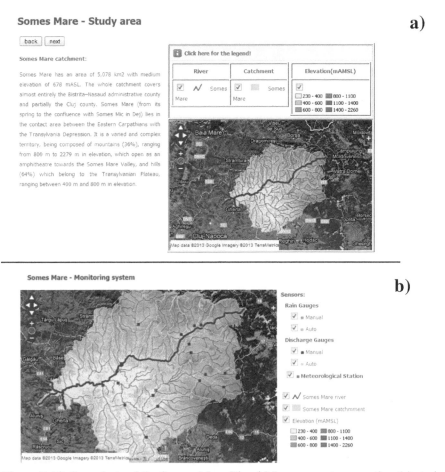

Figure 7.13. Snapshots of the Somes Mare Flood Management: presenting (a) study area and (b) monitoring station
Almoradie (2013)

Flood Information System

Flood information system (FIS) is the richest section of the platform in terms of provided information. It is divided in two parts aimed for citizens and professionals.

The section *Citizens' Flood information system* has two main elements. These elements are the *flood information explorer* and *citizens/stakeholder feedback* (shown in Figure 7.14).

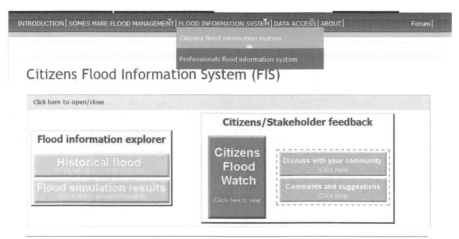

Figure 7.14. Citizens' Flood information system (Somes Mare)
Almoradie (2013)

The *flood information explorer* is intended to raises citizens' awareness about floods in different sections of the river. Historical floods with their description, cause and impacts are presented (see Figure 7.15) to guide citizens in exploring the historical data. Using an interactive map and chart interface, users can access monitoring station's time series of observed discharge, precipitation and temperature. With an aide of discharge thresholds, this provides visualisation if an event is high enough to cause flooding.

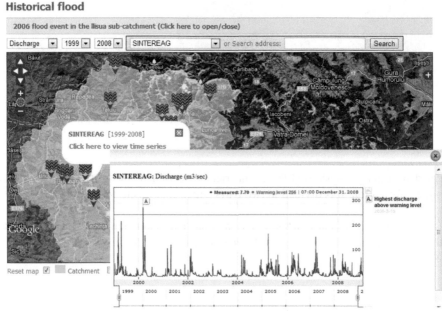

Figure 7.15. Citizens' FIS (Somes Mare)- Access to historical flood
Almoradie (2013)

Flood models are most often used as a supporting tool for long term FRM or for flood forecasting. To demonstrate the importance of models' for FRM model results of discharge time series (provided by the Somes Mare HEC-HMS model) and flood maps of 1/20 and 1/100 years return are presented in the flood information explorer (see Figure 7.16). The flood maps were generated separately and provided by Romanian Water Authority. This information will enable citizens and stakeholders to know the areas exposed to flood risk under these scenarios.

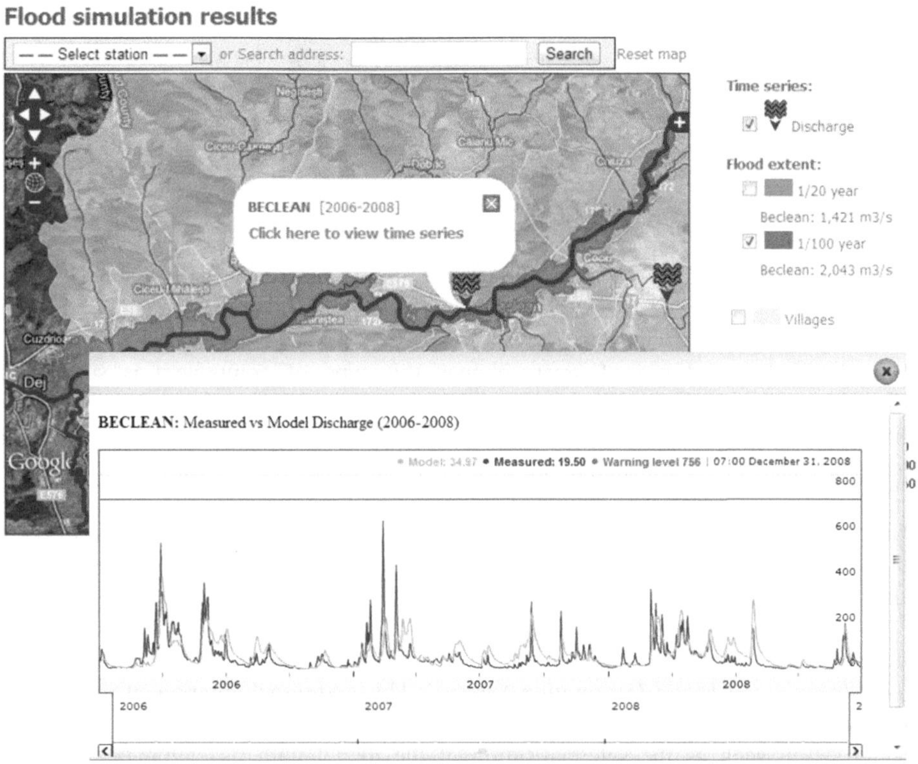

Figure 7.16. Citizens' FIS (Somes Mare)- Access to simulation results
Almoradie (2013)

The element *Citizens/stakeholders feedback* enables citizens and stakeholders to share information, or provide timely reports on local flooding. With the use of a forum they can discuss flood related issues with their community or even experts. The "*Citizens flood watch*" is the most interactive part of this element (Figure 7.17). One of the main functionality of this component is the use of location-based flood-related feedback for users. The user first provides the basic information (name, e-mail, address) and flood related textual information with supporting image (optional) in the input boxes, then he or she has to locate (by navigating or using search function) and click on the map the area where these flood related issues are. Once submitted the information is automatically published in the website (without the personal data).

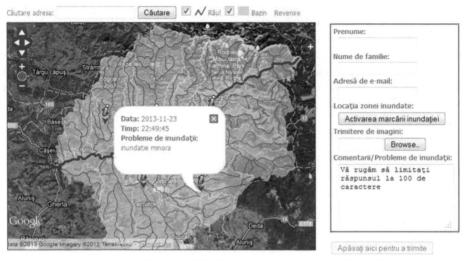

Figure 7.17. Citizens' FIS (Somes Mare)- Citizens Flood Watch
Almoradie (2013)

With the widespread use of smart phones this type of component is seen to be beneficial. Using the mobile phones web browser or a customised developed application user can send flood related information (in text or images) based from their current location. This is then published in the Citizens flood watch component.

Part of the Citizens/stakeholders feedback element is the discussion forum and feedback forms. The feedback forms provide users' the possibility to send suggestions and comments to improve the website or NESP flood platform.

Moreover, the "Flood Information System" has an additional component namely the *"The Professional Flood Information System"*. This additional component intended for professional users presents the HEC-HMS model of the Somes Mare catchment. The model set-up such as the schematisation of the model with its elements (sub-basin, reservoir, reach, etc) are presented via a map interface as shown on Figure 7.18.

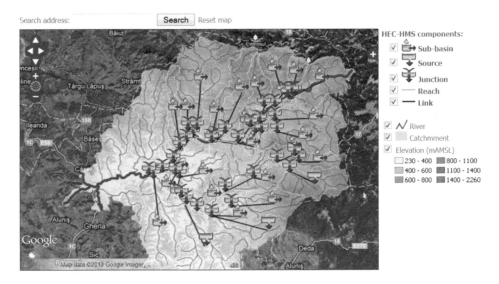

Figure 7.18. Professionals' FIS (Somes Mare)
Almoradie (2013)

Data Access

The "*Data Access*" is the last section of the Somes Mare NESP flood platform. This section provides users access to download time series and spatial data. The Data Access is most likely to be used by professional (e.g. modellers, decision makers) users rather than the citizens. This section is divided in two sub sections: "*Hydro-meteorological data*" and "*Spatial data*".

The sub section 'Hydro-meteorological data" presents an interface to access and download time series data in WaterML 2.0 format. Downloadable time series data are precipitation, discharge and temperature for the year 2007. In the interface users has to first select the type of data to download (precipitation, discharge or temperature) using the drop down box (in the map interface markers changes for different type of data). Users' then have to navigate the map and select the monitoring station by clicking on the marker. Once selection has been done, the data item is displayed in a box at the right side of the map; note that multiple selection of monitoring station and data is possible. Users can then view or download the data using the buttons in the box. Figure 7.19a shows 3 data items already selected (two discharge and one temperature data item). Data downloaded is in xml format (Type-of-data_station-name.xml – e.g. the second item from the list in Figure 7.19a would be Discharge_BECLEAN.xml). Data can be viewed as well directly via web browser, as shown in Figure 7.19b.

Hydrometeorological data access   a)

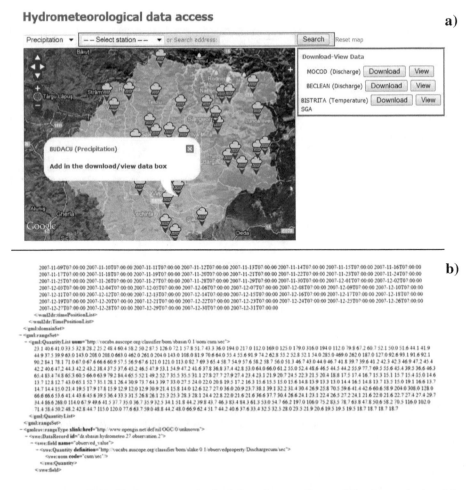

Figure 7.19. Hydro-meteorological data Access (Somes Mare): presenting (a) Interface for accessing times series data in WaterML 2.0 format and (b) Viewing time series data in WaterML 2.0 format
Jonoski et al. (2013) (Adopted)

In the "Spatial data access" sub section, users' can download and view spatial data such as river network, DEM, basin, etc. Spatial data can be downloaded in KML or XML format file. Using the WMS URL link, data can be viewed in the web browser. Figure 7.20 shows the river network accessed and viewed via a web browser using WMS URL link.

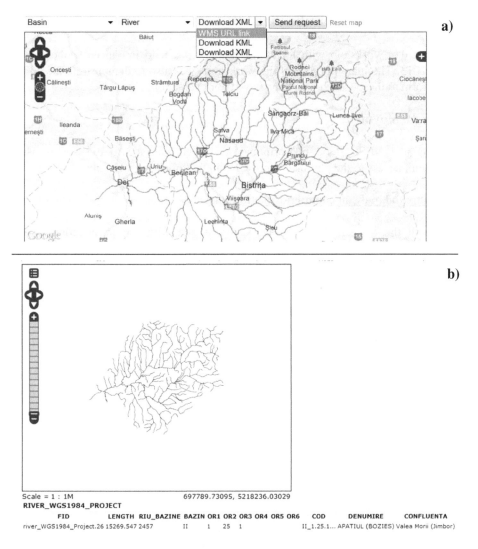

Figure 7.20. Spatial data Access (Somes Mare): presenting (a) Interface for accessing spatial data and (b) Viewing spatial data using WMS URL link
Jonoski et al. (2013) (Adopted)

7.4.2   Stakeholder evaluation

The evaluation by stakeholders for the Somes Mare case study is presented together with the Danube case study (see section 7.4.2) since both case studies have similar applications and evaluation questions.

7.5   NESP-CP 2: Danube river (Braila-Isaccea section)

A half a day workshop with local stakeholders was held in Tulcea at the end of February 2013. The main purpose of the workshops was also to demonstrate the NESP flood platform to these stakeholders and have their evaluation.

7.5.1 Deployment

The Danube NESP flood platform has a similar interface to the Somes Mare platform. Full access to the platform is provided in the following web link:

- http://hikm.ihe.nl/envirogrids/platform/danube/

Similar with the Somes Mare, the workshop started with the introduction of the platform, followed by a detailed presentation and demonstration of all platform functionalities. After this phase it was planned to have the participants test the NESP flood platform. However, due to unavailability of sufficient number of computers with internet connections hands-on test by participants was set aside. Instead of a hands-on test, questions regarding the use of the platform were asked from the participants, and these were answered with the use of the NESP flood platform. In the end the participants received evaluation forms with a set of evaluation questions that they were requested to answer. Generally the workshop went well.

The following presents a summary of the results from the NESP flood platform application.

Braila-Isaccea flood management

The "*Introduction*" and "*Study area*" section for the Braila-Isaccea Flood Management platform has similar presentation with the Somes Mare platform as shown in Figure 7.21.

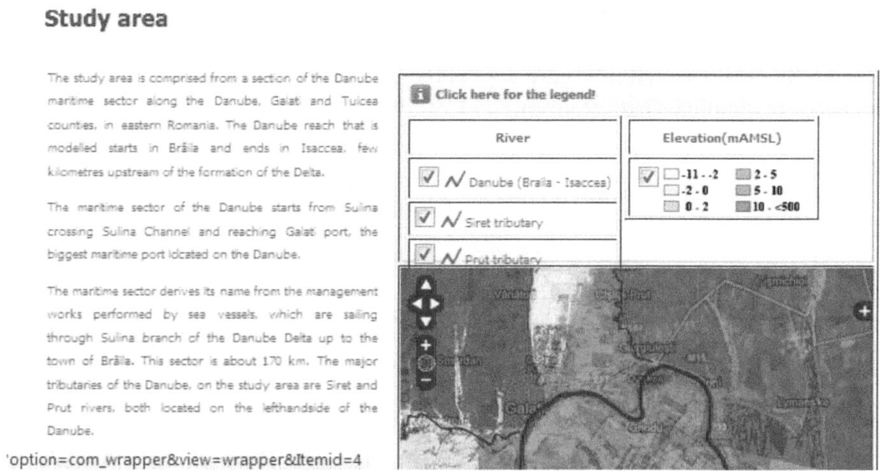

Study area
The study area is comprised from a section of the Danube maritime sector along the Danube, Galati and Tulcea counties, in eastern Romania. The Danube reach that is modelled starts in Brăila and ends in Isaccea, few kilometres upstream of the formation of the Delta.
The maritime sector of the Danube starts from Sulina crossing Sulina Channel and reaching Galati port, the biggest maritime port located on the Danube.
The maritime sector derives its name from the management works performed by sea vessels, which are sailing through Sulina branch of the Danube Delta up to the town of Brăila. This sector is about 170 km. The major tributaries of the Danube, on the study area are Siret and Prut rivers, both located on the lefthandside of the Danube.
'option=com_wrapper&view=wrapper&Itemid=4
Click here for the legend!
River
Danube (Braila - Isaccea)
Siret tributary
Prut tributary
Elevation(mAMSL)
-11 - -2
-2 - 0
0 - 2
2 - 5
5 - 10
10 - <500

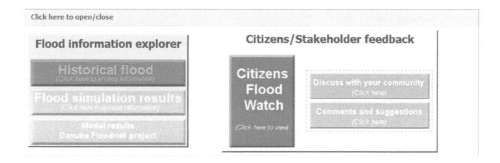

Figure 7.21. Snapshots of the Braila-Isaccea Flood Management: presenting study area
Almoradie (2013)

The sub-sections “Flood Problems”, “Flood Directive and Legislation”, “Current Management Plan” and “Basin Committee” similarly provides textual information combined with images and downloadable documentation.

Flood Information System

Unlike the Somes Mare, the Danube FIS was only aimed for the citizens'. The “*Flood Information System*” section has an interface to its subcomponents as shown on Figure 7.22.

Click here to open/close
Flood information explorer
Historical flood
(Click here to access information)
Flood simulation results
Model results
Danube Floodrisk project
Citizens/Stakeholder feedback
Citizens
Flood
Watch
(Click here to view)
Discuss with your community
(Click here)
Comments and suggestions
(Click here)

Figure 7.22. Citizens' Flood information system (Danube)
Almoradie (2013)

The components in the “*Citizens / Stakeholder feedback*” will not be presented here since it is similar with the Somes Mare platform.

The *"Historical flood"* component under the "Flood Information Explorer" presents historical discharges of the 3 monitoring stations in the Braila and Isaccea section of the Danube River. Presented in Figure 7.23 is the interface to access and view the historical discharge (example Grindu station).

Figure 7.23. Citizens' FIS (Danube)- Access to historical flood
Jonoski et al. (2013) (Adopted)

The "*Flood simulation results*" is the most interactive and interesting sub section of the "Flood Information Explorer". Flood maps (extent and flood depths) of the SOBEK 1D-2D modelling results are presented. Using a matrix interface users can interactively compare the results of the different scenarios over the base map by overlaying two different flood maps (in different colors), as shown in Figure 7.24.

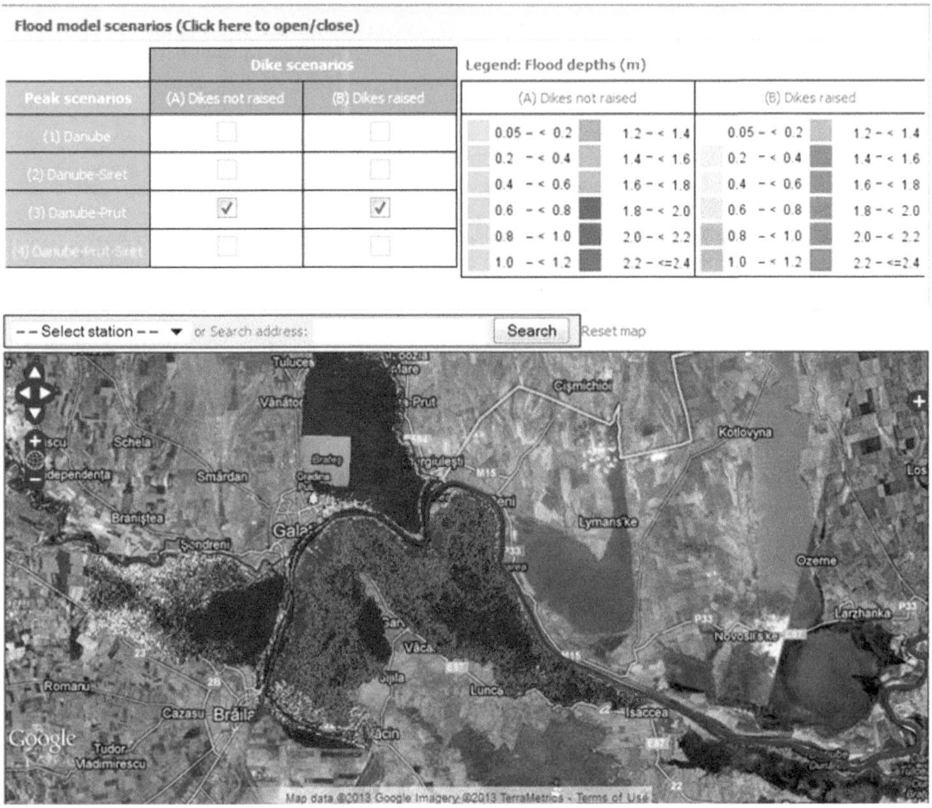

Figure 7.24. Citizens' FIS (Danube)- Visual comparison of the flood maps for composite scenarios modeled by the SOBEK 1D-2D model.
Almoradie (2013)

These flood maps can also be compared with the 1/1000 year return period flood hazard and flood risk maps obtained from an earlier research project (Danube FloodRisk EU project)- website shown in Figure 7.25. The estimated Danube discharges of this return period is 20% higher compared to the 1/100 year return period (about 18,500 m^3/s, compared to 15,500 m^3/s). Comparing the two return periods, the flooding pattern is very similar with differences in flood depths; 1/1000 year return period has a higher flood depths.

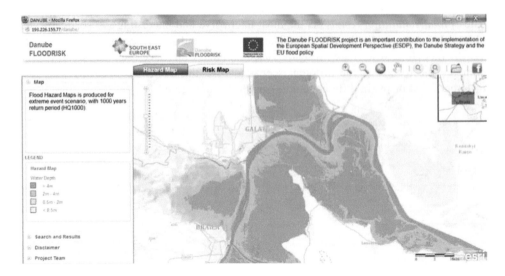

Figure 7.25. Citizens' FIS (Danube)- Flood maps for 1/1000 return period obtained from the Danube FloodRisk research project

Data Access

In the "*Data Access*" section downloading of time series data in the "*Hydro-meteorological data*" sub section is made available in Excel files. The "*Spatial data*" sub section is similar to the Somes Mare platform. The flood maps from the model results are also made available for users to download. Figure 7.26 presents an example of selecting and displaying a spatial data via the WMS URL link.

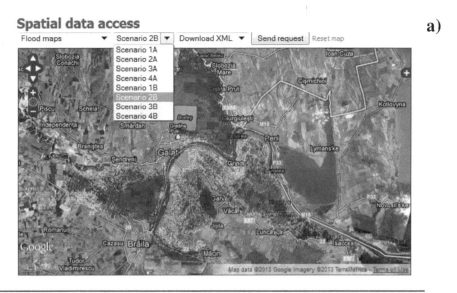

Scale = 1 : 487K
Click on the map to get feature info

Figure 7.26. Spatial data Access (Danube): presenting (a) Interface for accessing
spatial data and (b) Viewing spatial data (flood maps) using WMS URL link
Jonoski et al. (2013) (Adopted)

7.5.2 Stakeholder evaluation (Danube and Somes Mare)

Evaluation process

After the hands-on testing during the workshop (for both the Danube and Somes Mare platform) evaluation forms were handed out to the participants. The evaluation form contains structured set of evaluation questions for participants to answer. For the Somes Mare workshop a total of 21 participants carried out the evaluation and for the Danube workshop a total of 14 participants.

The evaluation questions for both case studies are the same. In the evaluation forms participants were first asked to provide information about which group of stakeholders they belong. These groups of stakeholders are the following:

- Government agencies and local councils
- Emergency agencies
- Companies and businesses
- Research institution
- General public
- Other

The main evaluation questions are as follows:

1. Is the structure of the flood platform clear and easy to follow?
2. Is the information presented in the web platform sufficient to raise awareness about flooding?
3. Please specify which type of information/application you find useful in the web platform?
4. Is the historical discharge with warning level informative and useful to raise awareness about local flooding?
5. Are the flood maps (model results) useful to raise awareness on the vulnerability of the areas if such flood scenarios happen?
6. Are the feedback components (Citizens flood watch, forum, comments/suggestions) useful   for citizen / stakeholder participation in flood risk management (FRM)?
7. Would you recommend this flood platform to your colleagues, family and friends?

For the first two questions participants answers the questions based on scales ranging between 1 (lowest score) and 5 (highest score). While linguistic descriptions were used for other questions (e.g. for question 5: not useful at all; not useful; neutral; useful and very useful).

Figure 7.27 presents the number and percentage of stakeholders group that participated in the workshop.

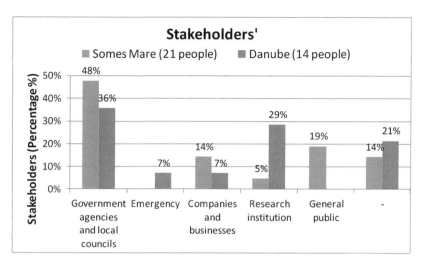

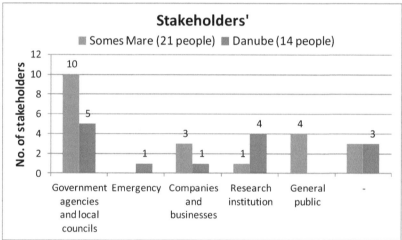

Figure 7.27. Summary of clustering of participating stakeholders (percentages and numbers) in the Danube and Somes Mare workshop
Almoradie (2013)

Most of the participants belonged to government agencies and local councils. However there are also a number of participants representing the other stakeholder groups.

Charts were used to visualise the evaluation by the stakeholders.

Evaluation results

Figure 7.28 presents the summary of participants responses on the main evaluation questions.

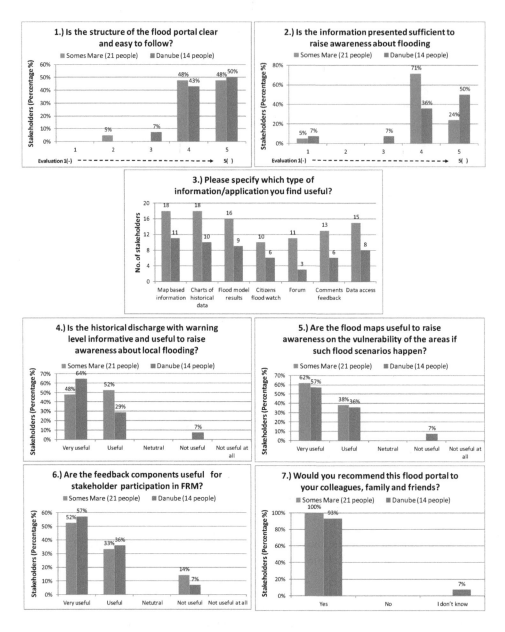

Figure 7.28. Results of the NESP flood platforms evaluation by the stakeholders
(Danube and Somes Mare platform)
Almoradie (2013)

As shown in the above figure, majority of the stakeholders find the structure of the portal easy to follow and with sufficient information to raise awareness on flooding. These can be attributed to the portals carefully designed structure, contents and interactivity.

The map-based information and charts of historical data were seen as the most informative and consequently most important applications. The citizens flood watch and the forum were seen as less useful.

Historical time series data with warning levels and the flood model maps were seen to be valuable for raising citizens' awareness on floods. Moreover, participants appreciated the value of the feedback component to facilitate stakeholder participation in FRM.

With the positive response from the participants, they would highly recommend the flood portal to their colleagues, friends and family to raise their level of awareness and knowledge on flooding issues in their local areas.

In the end of the evaluation questions participants also provided comments / suggestion about the platforms. The following is a combined summary of comments and suggestions for both platforms:

Stakeholders' comments and suggestions:

- It is recommended to add examples on how to use the system by specific users, such as mayors, city hall employees, environmental agencies;
- It is recommended to add the platform to different sites not just to enviroGRIDs site. Examples of sites where it can be linked is: city hall website, local newspaper, local weather site;
- We are looking forward to have the same platform for other catchments in the area;
- We are looking forward to have the same platform for the entire Somes basin;
- We would like the platform to be extended to other basins as well;
- This example should be extended to other branches of the Danube, in particular to Branch Maciu, where there were many dike breaches in the past. These kinds of phenomena triggered a lot of social problems. A second wish is to look at floods with smaller return periods as well, such as floods with return periods 1 in 20 years and 1 in 5 years.
- Is it possible that the ones who developed such a platform are implementing a programme which can verify the validity of the feedback given by the citizens and the frequency with which these data is provided to the platform?

As seen from the evaluation and comments, in general both platforms are very much appreciated and are seen as a valuable tool by the participating stakeholders. Recognizing its value, participating stakeholders from both case study recommended developing similar platform to other basin in Romania.

7.6   NESP-CDM 1: Cranbrook catchment

The structure, look and feel, contents and interactivity of the CP and CME of the Cranbrook platform was customised to satisfy the requirements of the case study. A summary of the results from the applications will be provided here, while the web links provided below give full access to both platforms.

7.6.1 Deployment

The Collaborative Platform

Figure 7.29a presents the interface to access the components of the CP. Figure 7.29b contains sample snapshots of the components in the CP.

Figure 7.29. Snapshots of the Collaborative Platform (Cranbrook case study).
Almoradie et al. (2013) (Adopted)

For hands-on experience CP can be accessed using the following web links:

- Cranbrook catchment: http://hikm.ihe.nl/diane_cm/cranbrook/index.php

The CP was extensively used for providing and managing relevant information during the workshops and it also served as a tool for interaction between the modellers and other stakeholders remotely in the periods between the workshops. The workshops were used by the modellers to present in detail the proposed alternatives, objectives and corresponding modelling results to the stakeholders and to learn more about the values and interest of the stakeholders. Furthermore, the face to face workshops were found to be very important since they initiated and maintained the interest of the stakeholders and contributed to building trust among all participants.

The relevant information presented in the CP are the (1) *Cranbrook study area, monitoring system, rainfall data and model description*, (2) *Flood risk framework- terms related to flood risk*, (3) *FRM framework-* flooding problems, legislation, current flood risk, objectives identification, external scenarios, alternatives and evaluation criteria and ranking of alternatives, and (4) the *Stakeholders*.

The *forum* (feedback features) of the platform was not intensively used. Most of the stakeholders preferred to give comments during the workshops using feedback sheets or via e-mail. The participants in the UK case study also preferred to use the online feedback forms rather than the interactive forum.

This first stage of collaboration was useful for reaching a shared understanding of flood risks and learning about the values / interest of the participating stakeholders. This led to the identification of new alternatives and the redefining of the objectives that were originally proposed by the experts. As a result a new set of proposed alternatives and objectives were defined together with the stakeholders, which were subsequently used for the CME.

Different alternatives for dealing with flood risk were initially proposed and later on discussed amongst stakeholders, based on the potential benefits and implementation feasibility. The final set of proposed alternatives together with the modelling approach used is presented in Table 7.1. The do nothing was included for comparison purposes amongst other alternatives.

Table 7.1. Alternatives for the Cranbrook catchment case study

Alternative	Description	Modelling approach
Alternative 1:	Do nothing (base case)	Base case 1D-2D model
Alternative 2:	Rainwater harvesting	Base case model with introduction of additional storages representing roof or garden rainfall harvesting
Alternative 3:	Improved and targeted maintenance regimes for the sewer	Base case model with and without clogging of strategic pipes

	system	
Alternative 4:	Improved resistance for preventing water from entering properties	Base case model with additional introduction of sandbags, floodsaxs and raised sidewalks
Alternative 5:	Improved rainfall and flood forecasting and warning	Base case model

The Objectives for the Cranbrook catchment are shown in Table 7.2, together with the used indicators.

The analysis of the performance of the alternatives was carried out by considering several flooding scenarios composed of the following factors:

- rainfall return period (30 or 200 years)
- water level at Roding River (low or high)
- profile of the rainfall storm (winter or summer profile, from the UK Flood estimation handbook, Centre for Ecology and Hydrology (1999)

Scenarios with different combinations of these factors were modelled by the project partners (Imperial College of London). In the final analysis the most critical scenario of 200 years rainfall return period, high water level at River Roding and summer profile of the rainfall storm was used for testing the performance of the alternatives. The stakeholders were made to understand that the probability of coincidental occurrence of these scenarios is highly uncertain (rarer than 30-200 years). Although this is highly uncertain to occur, experts and stakeholders prefer this critical scenario because it has the most significant flooding.

Table 7.2. Objectives for the Cranbrook catchment case study

	Obj 1	Obj 2	Obj 3	Obj 4	Obj 5
Objective (Obj.)	To reduce the magnitude of surface flooding	To minimise the damage to properties	To minimise damage to critical infrastructure	To maximise the opportunity of salvaging belongings	To maximise ease and feasibility of implementatio n
Indicator	Flooded area where flood depth is above 30 cm. (Units: flooded hectares)	Number of properties flooded	How would you rate the flood damage to critical infrastructure?	How would you rate the opportunity to salvage valuables inside properties and businesses from flood damage?	How would you assess the ease and feasibility of implementing this measure in Redbridge?

| Type of indicator / scale for evaluation | Quantitative indicator, the evaluation is based on the results of the flood model and the user cannot modify these results. | Quantitative indicator, the evaluation is based on the results of the flood model and the user cannot modify these results. | - Very Low damage
- Low damage
- Medium damage
- High damage
- Very High damage | - Very Low opportunity
- Low opportunity
- Medium opportunity
- High opportunity
- Very High opportunity | - Very Low feasibility
- Low feasibility
- Medium feasibility
- High feasibility
- Very High feasibility |

As can be seen from the tables above the quantitative indicators for Objectives 1 and 2 would be same for all stakeholders because they are simply read as modelling results. For the remaining three objectives the stakeholders are provided with extensive information on the platform (e.g. maps of critical infrastructure or exposed properties merged with flood inundation maps, characteristics of a future flood forecasting and warning system, such as expected lead time, etc), however each individual stakeholder can still provide different evaluation using the linguistic terms. Together with the fact that different stakeholders also provide different weights to the different objectives this leads to different individual rankings of the proposed alternatives in CME.

The Collaborative Modelling Exercise

Figure 7.30 presents the interface to access the components of the CME.

Figure 7.30. Interface to access the components of the CME (Cranbrook case study).
Almoradie (2013)

The CME can be accessed using the following web links:

- Cranbrook catchment: http://hikm.ihe.nl/diane_cm/internal/cranbrook/exercise

The second stage of collaboration using the CME started with a face to face workshop and afterwards the stakeholders were given a month to use the platform for remote interactions and negotiations. In the Cranbrook case study eight participants took part in the CME. The workshop started with the introduction of the platform, presenting them the objective of the platform and the steps involved in developing and using the Individual and Group profiles. After this introduction, stakeholders were given about an hour to have hands-on experience with the CME platform, while the developers and experts/modellers were guiding and supporting them. Afterwards the stakeholders continued using the platform rather independently. In general this hands-on exercise went smoothly, however there were clarifications needed in some parts of the individual and the group profile, related to the usage of the interface and clarification of the graphical presentation of the obtained results. Some participants were not very familiar in using the Internet or a computer, thus developers and modellers needed more time in guiding them. The actual negotiation stage was never reached. The participants recognised the value of using the platform in the intended way, but without the urgency to engage in FRM activities it was difficult to maintain continuous engagement of the stakeholders beyond the workshops dedicated to CME. Some stakeholders used the additional month to update their individual profiles online, but without engaging in negotiations with others.

As presented in the conceptual framework in Chapter 4 this part of the NESP application (Collaborative Modelling Exercise) has two main components, the "*Individual profile*" and "*Group profile*". Figure 7.31 shows the workflow of the individual and group profile. This is followed by the implementation and results of each component.

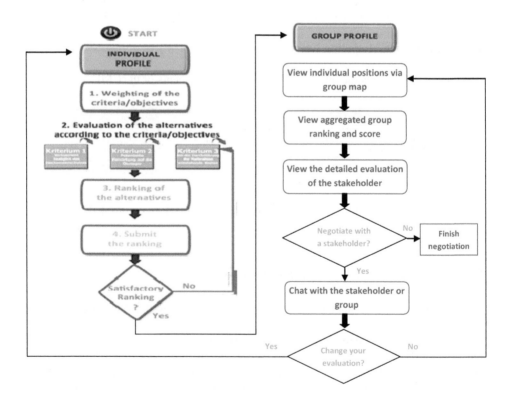

Figure 7.31. Workflow from individual to the group profile
Almoradie et al. (2013) (Adopted)

Individual profile

Some impression about the steps that an individual stakeholder takes during the evaluation of the proposed alternatives is provided via Figures 7.32 and 7.33. Figure 7.32a presents the initial interface in CME through which stakeholders can access the following components: providing weights to the objectives, evaluation of alternatives (the workflow is per objective), obtaining the ranking and submitting the ranking. The evaluation of alternatives for a selected objective is carried out via interfaces shown on Figure 7.32b. Map-based information guides the user to provide the evaluations in linguistic terms. Once evaluations for all alternatives are finished the user is presented with a summary decision matrix (Figure 7.33a) from which the final ranking can be obtained (using TOPSIS) and plotted as on Figure 7.33b. In addition to the individual ranking of alternatives, which is the key component of the individual profile, the stakeholders are also asked to provide some additional information to complete their profile. This information consists of textual explanation about the reasoning behind their evaluation, indication about their perception on responsibilities of different agencies for implementing the proposed alternatives and possible suggestions for new alternatives that may need to be tested. The complete individual profile is subsequently stored in a database on the server so that it can be made visible upon request to other stakeholders.

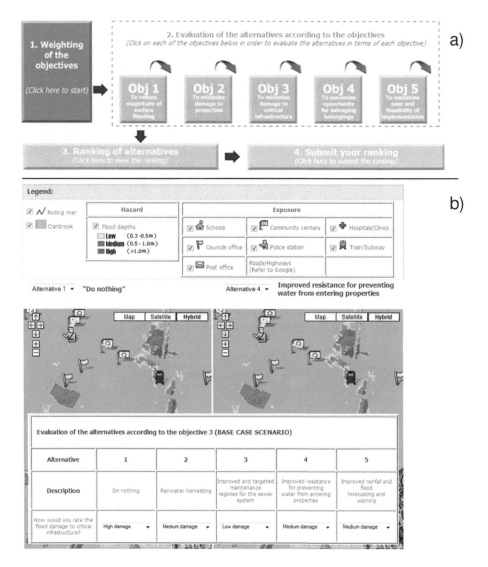

Figure 7.32. Interface elements for evaluation of alternatives (Cranbrook case study):
(a) Interface through which stakeholders can provide weights to the objectives, evaluate
alternatives and obtain ranking. (b) Interface for evaluating the alternatives for a
selected objective; a map-based interface guides the users to provide the evaluation in
linguistic form.
Almoradie et al. (2013) (Adopted)

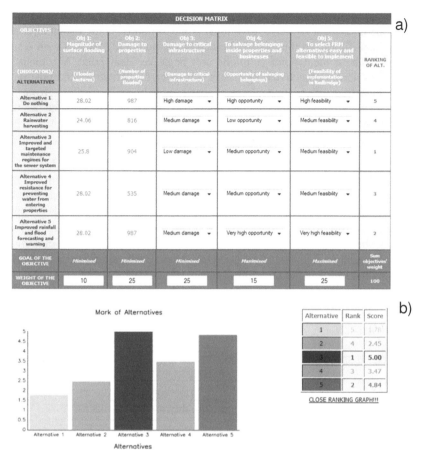

Figure 7.33. Decision matrix and individual ranking (Cranbrook case study)
Almoradie et al. (2013) (Adopted)

The following provides some more details about the methods behind the individual ranking.

In order to obtain the individual ranking an individual stakeholders needs to carry out two critical steps: (1) provide weights to the identified objectives (by distributing points from a maximum of 100), by means of which the relative importance of the objectives is specified, and (2) evaluate the alternatives according to the objectives. Evaluations are with quantitative or qualitative indicators for the identified objectives. Quantitative evaluation is based on the results of the hydrological or hydraulic models, thus users cannot modify these. Qualitative evaluations are expressed in linguistic terms (e.g. bad, fair, and good). Stakeholders evaluate the alternatives by choosing from these linguistic terms based on their point of view on how certain alternative performs. Since the qualitative evaluation is expressed in linguistic terms, these are converted into crisp numbers by using conversion scales based on fuzzy set theory in MADM. (see Chen et al. (1992) for details about the conversion scales).

With the so-determined evaluation outcomes for all alternatives the TOPSIS method is applied for the individual ranking. Two ideal point solutions (non-existent alternatives) are initially identified from the available outcomes from all alternatives with respect to all objectives. These ideal point solutions are: negative ideal point A⁻, consisting of worst outcomes across objectives regardless from which alternative, and positive ideal point A⁺ consisting of best outcomes across objectives regardless from which alternative. The score for each actual alternative D_i is then calculated as relative distance to these ideal solutions:

$$D_i = \frac{S_i^-}{S_i^+ + S_i^-} \tag{1}$$

, where S_i^- and S_i^+ are Euclidean distances between the actual alternative (i) and the negative and positive ideal solutions respectively. From this score the rank of the alternative is obtained (highest rank for highest score). In addition to the ranking, these scores are also used to calculate the relative closeness of the alternatives between each other by using the following expressions:

$$CR_i = \frac{D_i}{D_{max-ranked}} \tag{2}$$

, where CR_i is named 'closeness ratio' and $D_{max-ranked}$ is the score for the best alternative.

The so-called Borda score BS is calculated from the ranking results as:

$$BS_i = N + 1 - Rank_i \tag{3}$$

, where N is total number of alternatives and $Rank_i$ is the rank of the given alternative.

Using the following expression for the relative position RP_i of a given alternative in-between the alternatives ranked above and below it:

$$RP_i = \frac{CR_i - CR_{lower-rank}}{CR_{upper-rank} - CR_{lower-rank}} \tag{4}$$

, a final score FS_i is calculated as :

$$FS_i = BS_i + RP_i \tag{5}$$

These final scores are subsequently used for the aggregations in the group profile and in additional plots for presenting individual positions versus the group.

Group profile

The group profile aggregates the final scores of all stakeholders and provides the results in graphs and tables. Special visualisation components are provided for comparing the individual positions of the stakeholders within the group and access their individual profiles.

It is important to note that the rankings provided with this procedure are not at all considered to be final, but rather a starting point for subsequent collaboration and negotiation. Following negotiations an individual stakeholder may decide to update his or her individual ranking, which would immediately update the group profile as well, so that further progress in the negotiations can continue. After many rounds of exploratory and progressive use of the platform, the negotiations may conclude with possible recommended alternatives for implementation. This is the envisioned usage of the platform, even though, as will be shown later, such progress in negotiations was not reached in this first implementation of the platform.

Three different visualisations were used to present the final results, shown in Figure 7.34. Figure 7.34a presents a special visualisation entitled 'swimming pool of alternatives' (because of the usage of the different shades of blue colours). In this visualisation more preferred alternatives by the whole group are indicated by darker blue colours and how individuals rank the same alternatives is indicated by individual markers (clustered in yellow when there are more individuals with same position). The "balloon" markers are the positions of individual stakeholder; the colours correspond to his/her stakeholder group. The "human" marker presents the position of the current user. Clicking the markers the individual profile (including the ranking) of any participating stakeholder become visible to all other participants. Same results are also available in additional tables and graphs. Figure 7.34b shows the aggregated group ranking in bar charts. Figure 7.34c summarises the individual raking of each participant in a table and also displays the group score for each alternative (the sum of individual scores).

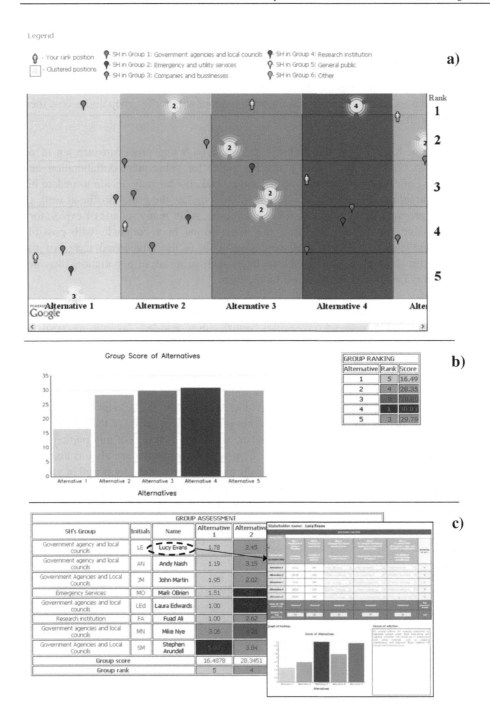

Figure 7.34. CME results for Cranbrook
Almoradie (2013)

The results show that the all alternatives (except alternative 1- 'Do nothing') had similar overall ranking with slight preference for Alternative 4 ('Improved resistance for preventing water from entering properties').

These results are certainly not conclusive. The number of participating stakeholders was quite small and the CME did not proceed further with active negotiations. Nevertheless they give first indication about the attitudes of the actual stakeholders in the case study towards the proposed alternatives and can serve as a starting point for further collaboration in deciding which alternatives may be commonly accepted by the respective communities.

Although negotiations did not go further, for future use the platform is equipped with chat functionality to support remote negotiation between stakeholders, shown in Figure 7.35.

Figure 7.35. CME Chat interface
Almoradie (2013)

7.6.2 Stakeholder evaluation

The evaluation of stakeholders for the Cranbrook case study are presented together with the Alster case study (see section 7.6.2) since both case studies have similar evaluation questions.

7.7   NESP- CDM 2: Alster catchment

The structure, look and feel and processes of the Alster CP and CME is similar to the Cranbrook case study. However the platform was customised to satisfy the case study's requirements. The platform was implemented in German language. The following section provides a summary of the results from the applications. Web links are provided below to give full access to the platforms.

7.7.1 Deployment

The Collaborative Platform

Figure 7.36a presents the interface to access the components of the CP. Figure 7.36b contains sample snapshots of the components in the CP.

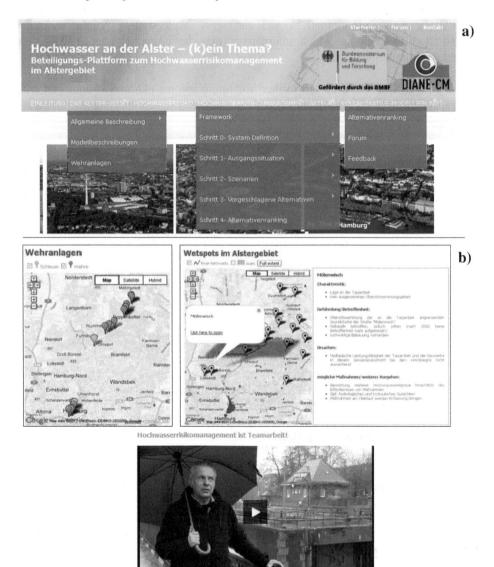

Figure 7.36. Snapshots of the Collaborative Platform (Alster case study).
Almoradie (2013)

The CP can be accessed using the following web link:

- Alster catchment: http://hikm.ihe.nl/diane_cm/alster/index.php

Same as the Cranbrook case study the Alster CP was extensively used for the workshop as an information management-dissemination tool and as a tool for remote interaction between the modellers and other stakeholders in the periods between the workshops. Face to face workshops were used by the modellers to present in detail the scenarios, modelling results, proposed alternatives and objectives to the stakeholders. The workshops were also used to have a common understanding of issues and terms, and to learn more about the values and interests of the stakeholders. To learn more about their interest and local knowledge, selected stakeholders were video interviewed. These video interviews were then embedded in the CP.

The forum (feedback features) of the platform was not also intensively used. The use of feedback sheets or via e-mail was preferred by most of the stakeholders. However, there was a period during a flood event in February 2011 when the stakeholders made used of the forum. This shows that for certain situations forums can be useful.

As mentioned earlier, this first stage of collaboration was useful for reaching a shared understanding of flood risks and learning about the values / interest of the participating stakeholders leading to the identification of new sets of alternatives and objectives that were originally proposed by the experts. These new set of proposed alternatives and objectives, defined together with the stakeholders, were then used for the CME.

The collaboration during the first stage took somewhat different directions. During the workshops the focus progressively shifted from the originally intended catchment-based approach to FRM to identification of several critical points along the Alster river (so-called 'wet-spots') that have been frequently flooded in the past. Considering that the risk of flooding in the Alster is higher in the upper part than in the lower part, experts and modellers saw the importance of identifying these wet-spots; these were identified together with LSBG and stakeholders. Figure 7.37 presents the interface to access wet-spot information.

Eine Übersicht über die identifizierten Wetspots finden Sie hier

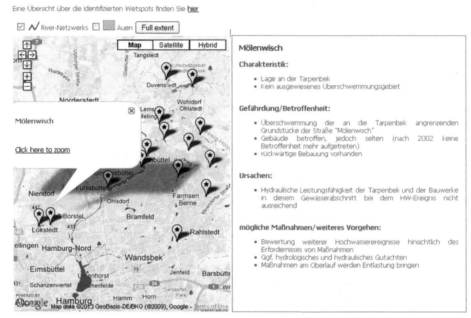

Figure 7.37. CP interface to access 'Wet-Spots' information
Almoradie (2013)

Nevertheless it was agreed that in CME the catchment perspective should still be maintained, with more general objectives and alternatives even without having modelling results on that scale. Consequently the alternatives were formulated as sets of possible measures (the effects of which could be visualised from modelling results in some particular wet spots) and the objectives were formulated with qualitative indicators. The alternatives and objectives for the Alster case are shown in Tables 7.3 and 7.4 respectively.

Table 7.3. Alternatives for the Alster catchment case study

Alternative	Description / some possible measures (full list available on the platform)
Alternative 1: Doing nothing (base case)	No measures are implemented
Alternative 2: Technical measures	Control of hydraulic structures; raising of crest levels at certain locations; construction of storage areas, etc...
Alternative 3: Activities in the catchment area / area management	Sustainable and careful maintenance of water systems (removal of obstacles), wetlands in the upper catchment area, etc...
Alternative 4: Preventive measures	Improved coordination; (private) property protection; forecast / information, etc...

Table 7.4. Objectives for the Alster catchment case study

	Criterion 1/ Obj 1	Criterion 2/ Obj 2	Criterion 3/ Obj 3
Criteria	Effectiveness for flood protection	Potential impact to the ecology	Cost of implementation of the alternatives
Accounted objectives	- To reduce the magnitude of surface flooding - To minimise damage to properties	- To increase the protection and enhancement of the floodplain	- To minimise the cost of implementation of the alternative
Indicator	How would you rate the effectiveness of the alternative in terms of flood protection to properties and infrastructure?	How would you rate the potential impact for ecology (protection and enhancement of the floodplain) with the implementation of each alternative?	How would you rate the total cost of implementation of the considered measures grouped in one alternative all along the identified wetspots? (This assessment was provided by LSGB and users cannot modify them)
Type of indicator / scale for evaluation	- Very Low effectiveness - Low effectiveness - Medium effectiveness - High effectiveness - Very High effectiveness	- Very bad impact - Bad impact - Neutral impact - Good impact - Very Good impact	- Very low cost - Low cost - Medium cost - High cost - Very High cost

The measures that were modelled were with consideration of several different scenarios mainly related to the flood return period. After considering 100, 200 and 300 year return period the 200 return period was used as a base case. Some extreme scenarios in the city of Hamburg associated with failure of tidal protection structures in combination with 200 year return period were also modelled, but they were not used in CME.

Similarly to the UK case study the usage of qualitative indicators and different weights for the objectives leads to different individual rankings in the CME.

The Collaborative Modelling Exercise

The CME can be accessed using the following web link:

- Alster catchment: http://hikm.ihe.nl/diane_cm/internal/alster/exercise/

Twelve stakeholders participated in the second stage of collaboration. Similar to the UK case study, it started with a face to face workshop using the CME. They were then given a month to use the platform for remote interactions and negotiations. The workshop first introduced the platform and its objective, followed by a brief demonstration in using the Individual and Group profiles. Stakeholders were then given about an hour to have hands-on experience with the CME platform with the support and guidance of the experts/modellers. Afterwards, stakeholders independently continued to use the platform for another hour.

Similarly to the UK case study, developers and modellers needed more time to guide participants not familiar in using the computer or the internet. However in general, the CME went well. Actual negotiation stage was also never reached despite the fact that they were given additional month to update their individual profiles online. However, the technical partner (LSBG) in the Alster case study expressed particular interest in the developed web tools with the intention of further development and adoption as a supporting tool for their future FRM activities.

This case study also has the *Individual* and *Group* profile as the main components of the CME- same as the Cranbrook case study. This is presented in the following.

Individual profile

Sample snapshots of the processes for building the individual profile are presented in Figures 7.38 and 7.39. Figure 7.38a presents the initial interface in CME through which stakeholders can provide weights to the objectives, evaluate alternatives (per objective), obtain the ranking and submit the ranking. Figure 7.38b is the interface for stakeholders to evaluate alternatives for a selected objective; a map-based interface is used to guide users in their evaluation. Once stakeholders finish their evaluations for all alternatives a summary decision matrix (Figure 7.39a) is made available, where they can still make some adjustment in there evaluation. Through this decision matrix they can obtain the final ranking using the same TOPSIS implemented in UK case study. Figure 7.39b shows this final ranking. This final ranking and all related data and information are then submitted to the database on the server for storage.

Please refer to the Cranbrook case study for the decision method used- TOPSIS and its implementation.

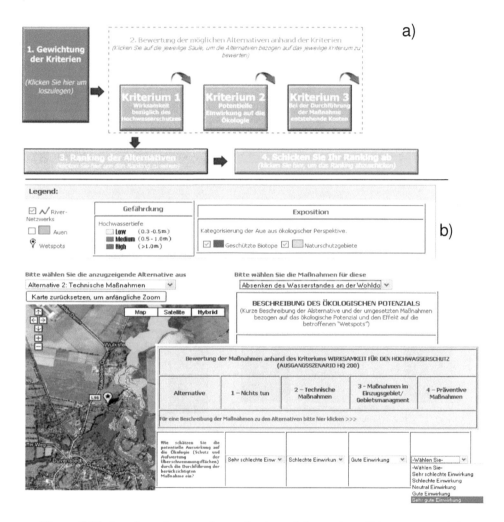

Figure 7.38. Interface elements for evaluation of alternatives (Alster case study)
Almoradie (2013)

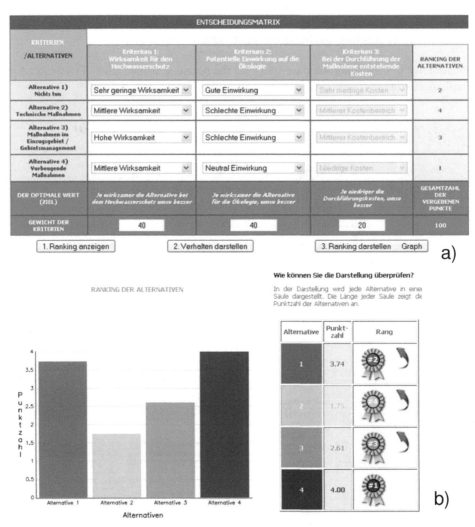

Figure 7.39. Decision matrix and individual ranking (Alster case study)
Almoradie (2013)

Group profile

The final results of the CME are presented in Figure 7.40. Figure 7.40a presents the swimming pool of alternatives, Figure 7.40b is the group ranking in bar chart and Figure 40c is the individual rankings summarised in a table. The same explanations of these visualisations as in the UK case study are applicable here.

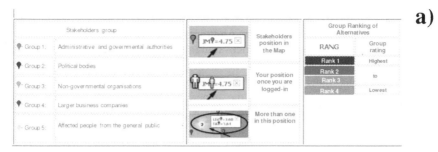

a)

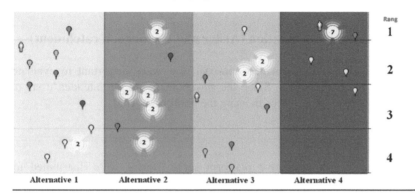

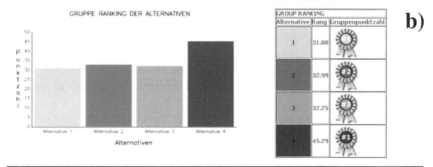

b)

c)

Figure 7.40. CME results for Alster

Almoradie (2013)

For Alster there is clearer preference for Alternative 4 ('Preventive measures') with a large cluster of 8 participating stakeholders ranking it as first. There is distribution of the positions of individual stakeholders across the proposed alternatives, often depending on the stakeholder category to which the individuals belong. For instance, in the Alster case, affected people from the general public generally ranked the 'Do nothing' alternative as last, while authorities and non-governmental organizations ranked the same alternative as second.

The results of Alster case study are also not conclusive since few participants were involved. Future applications should involve more and wider range of stakeholders.

7.7.2   Stakeholder evaluation (Alster and Cranbrook catchment)

Given the novelty of implemented approach it is very important to evaluate the developed collaborative platform. Since the stakeholders are the intended users of the platform, knowing their experiences in using the platform is vital.

Evaluation process

For evaluation of the collaborative platform survey forms were distributed to the stakeholders immediately after the collaborative modelling exercise. Same evaluation questions were formulated for both case studies as follows:

1. Is the structure of the exercise clear?
2. Is the information provided on the web-platform enough to complete the exercise?
3. Please specify which type of information you find useful on the exercise
4. Is the presentation of the flood maps informative and useful to support the definition of your preferences?
5. Is the individual ranking presented close to your representation of your preferences
6. Is the group view clear enough to support a negotiation stage?
7. Would you recommend the exercise to your colleague?

In the German case study, out of twelve stakeholders who participated in the CME ten answered the questionnaire and for the UK case study six people answered the questionnaire out of eight participants.

Evaluation results

Figure 7.41 provides a summary of the obtained results.

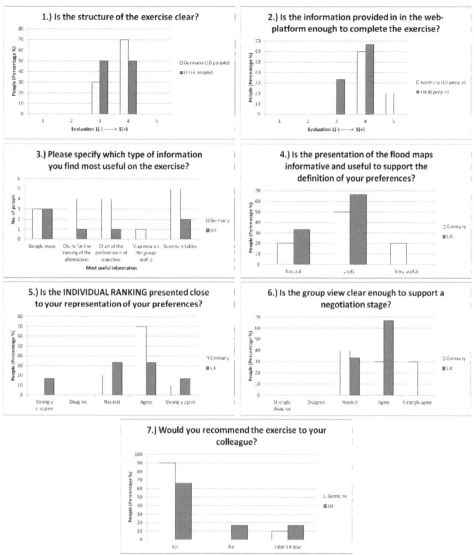

Figure 7.41. Evaluation of the CP and CME by the stakeholders (Alster and Cranbrook case study)
Almoradie et al. (2013) (Adopted)

Based from the answers of the stakeholders, it seems that the structure of the exercise is clear to them and the information provided on the platform is sufficient for them to complete the exercise.

The information that the stakeholders find most useful is in the summary tables and Google maps, while the map view of the group profile (the 'swimming pool of alternatives') is found to be less useful. The reason for this may be in the fact that this was an unconventional way of presenting the information and if the stakeholder could spend more time and become more familiar with it perhaps they would find this visualisation more useful. The flood maps were very useful and informative to support the definition of their preferences.

Most stakeholders agree that the platform provided a close presentation of their preferences regarding the measures and alternatives. The group profile was found to be clear enough to support a negotiation stage (even if this was not actually tested). Lastly the evaluation shows that a large majority of the stakeholders would recommend the exercise to their colleagues.

Stakeholders' preferences were somewhat influenced during the identification of possible alternatives and the individual and group assessment ranking of alternatives. Some of the stakeholders changed their perception and understanding of some of the measures. This was observed during the face to face workshops. Preferences also seemed to converge during the collaboration. For example in the Alster case, we can see in the group pool of alternatives (Figure 7.41) that the positions of individual stakeholders regarding alternative 4 (most preferred alternative) are much closer to each other.

The CME in general was appreciated by the stakeholders. Some important quotes from the stakeholders are:

" In order for it to be intensively used in the future, we would need to create a "culture" of using this platform".

"It would be interesting to have more people involved in order to have a broader view"

"The exercise should show more alternatives, more alternatives should have been modelled"

7.8   Concluding remarks

Here we summarize the conclusions for the different NESPs applied to different case studies.

7.8.1   NESP-IKS

The presented NESP-IKS demonstrates the benefits of using mobile phone and web applications for participatory water quality information communication between authorities responsible for environmental water quality and citizens as end users. The presented applications and its implementation showed innovative and new ways of communicating and sharing water- environmental related issues amongst citizens and stakeholders. Moreover, these applications may develop new kinds of partnership relations with shared benefits.

The developed web and mobile applications has been successfully tested on its design, functionality and usability. These are reflected from the participants (public and professionals) response on the evaluation questionnaires. In general the participants appreciate and value the importance of such tools or application. However there are still parts of the application that can be improve in its future development.

In the developers and water experts point of view, Android platform based smart phones can be used successfully to developed applications for water resources and flood management. In this study only a few of Android platforms numerous functionalities were tested. Location based (GPS and compass) and other sophisticated functionalities can be integrated in future development of such kind of applications. This can be developed as well in different water related fields of application (e.g. floods and emergency management, drinking water quality and quantity, and other environmental problems).

As demonstrated in this study mobile phone application is a promising tool to communicate water related information to stakeholders and citizens. The advancement and sophistication of this smart phones is unprecedented, hence the field of water resources and flood management should make use of its great potential not only for information dissemination but also for participation.

7.8.2 NESP-CP

The NESP-CP developed application is seen as a valuable system for awareness raising and citizens empowerment in FRM. Responses from diverse stakeholders during the workshop and on the evaluation questionnaires show that they highly appreciate and value such platforms. Authorities and stakeholders recognised the need of using such platforms for sharing of flood related information and knowledge. Further, they see the benefits of such platform in the management of floods. With such positive response, stakeholders and authorities would like to extend or developed such application for other basins.

The use of GPL and latest web technologies such as web services for publishing time series and spatial data has been successfully demonstrated in this case study. The use of standards (WMS, WFS, WaterML 2.0) for publishing and sharing time series and spatial data is seen to be really useful for experts, authorities and decision makers because of its interoperability. Moreover, knowing its advantages and benefits, European scientific communities are encouraging institutes, water authorities and other scientific communities to make use of such standards.

In general it can be concluded that while the NESP flood platforms have been successful as demonstrator applications, there is clearly a need for continuing further research and development of such applications that can lead to their broader adoption in the flood management practice.

7.8.3 NESP-CDM

The work presented demonstrates that web-based stakeholder collaboration in flood risk management based upon the NESP-CDM framework has potential to overcome the hindrances in stakeholder participation. The framework and the platform demonstrated to be useful in promoting interaction between stakeholders that integrates the collaborative modelling into the decision making process. Furthermore, the framework showed to be suitable in different contexts (local scale / pluvial flooding in Cranbrook and catchment scale / fluvial flooding in Alster). The use of German language for the Alster platform was necessary to have a good communication with the stakeholders. For stakeholders to gain interest, the platform structure and information should be clear enough for the stakeholders. Moreover, knowledge dissemination and information sharing prior to the use of the collaboration platform is important, which in these cases was achieved during the initial workshops.

Based from the overall feedback from the stakeholders it can be concluded that they recognise the usefulness of these platforms, however creating a "culture" of using these platform is clearly needed for their more intensive usage in the future. Such "culture" of using the platforms may possibly start when stakeholders realise the urgency of managing flood risks, which may provide the needed motivation for the stakeholders. The lack of such motivation may be the reason why actual negotiations stages were not reached during this research. If actual negotiation stages were reached the stakeholders would have to go through an iterative process of negotiating using the platform. They may use the forum and instant messaging components to negotiate. Along the negotiations, stakeholders may have to change their individual evaluation of an alternative several times. Using the 'swimming pool of alternatives' stakeholders can keep track on how close are their positions to those of the other participants, which may be supportive for converging to a possible alternative for implementation. In Europe, with the impetus created by the Flood Directive of the European Union, which requires development and implementation of FRM plans the CP and CME should be integrated in a longer participatory process and tested with a wider range of stakeholders over longer time periods. The fact that technical partner (LSBG) in the Alster case study expressed particular interest in the developed web tools and is looking forward for its further development is an indication that such developments may indeed follow.

One issue that was not fully addressed in the presented case studies is related to limitations of the used models and the associated uncertainty in their results. Elaborate uncertainty analysis that attempts to quantify these uncertainties was not carried out in any of the case studies. Although such analyses can be of value, it is still difficult to communicate its results to non-expert stakeholders. This is certainly a challenge for future similar studies. As an alternative approach to handle and communicate uncertainties, in this study (especially in the Alster case) stakeholders (expert and non-expert) were engaged to validate the model results by comparing them to their knowledge about recent flooding in local areas of the catchment.

Cost criterion (e.g. flood damage cost) was considered as one of the indicators for evaluation of an alternative. However since the quality and resolution of these data were not of sufficient quality, the final decision was not to include it in the final analysis. In the Cranbrook case study cost was instead replaced by other indicators (e.g. quantitative- number of properties flooded or qualitative-very low, medium, very high). In the Alster case study indicative cost assessments were included by using linguistic terms (high, low, medium) provided by LSBG, without giving the possibility to the broader stakeholders to modify these assessments. For future studies, if cost can be quantified properly, cost should be considered as one of the major indicators.

In the Judgement engine, the TOPSIS method provided a close representation of the stakeholders preferences regarding the measures and alternatives. However, for further research it will be of interest to look at other Multi Criteria Decision Methods.

The novel visualisation named 'swimming pool of alternatives' for the group view seems to be less important as a means of visual information. Stakeholders may have found it less useful because of its unconventional way of presenting the information. If stakeholders spend more time and become more familiar with it, and especially if they engage in negotiations this kind of visualisation may be found to be more useful. Nevertheless, other methods of visualisation for achieving similar goals need to be also explored.

Chapter 8
Conclusions and Recommendations

This research explores networked environment for stakeholder participation in water resources and flood management. This chapter presents conclusions drawn from this research and recommendations for future research and development of NESPs.

8.1 Conclusions

8.1.1 General conclusion

Networked Environments for Stakeholder Participation (NESP) are web-based computer-aided or mobile environments for remote virtual interaction between participating entities such as stakeholders. NESP is envisioned as an innovative approach in addressing the potential pitfalls or challenges-hindrances for stakeholder participation in water resources and flood management. Thus it can (1) Bring all together the stakeholders in an environment to participate in real time regardless of their spatial distribution (2) It cuts the cost of bringing all together the stakeholders (3) It can be a tool to provide innovative participation that could gain and maintain the stakeholders' interest and their long term commitment.

The work presented in this research demonstrated that NESPs such as web-based and mobile environments have potential to overcome the hindrances in stakeholder participation in water resources and flood management. The overall advantages of using such system are in reduced costs and increased efficiency of participation, while facilitating the sharing of common interests and objectives among citizens and different stakeholder. In this study, face-to-face workshops were needed to introduce and demonstrate these environments to potential users'. The number of face to face workshops depended on the type or level of participation. The use of case studies local language for the platforms was necessary to have a good communication with the stakeholders.

Iterative consultations with partner institutions or stakeholders were part of the process in finalising the design of the platforms-applications. A good example is the design of the Noord-Brabant NESP-IKS. Users were engaged (mainly professional users) continuously in the development of the environment to include their needs and requirements in the design and functionality. This made the potential of the application to be used in practice higher. Moreover, through this approach the platforms were seen as more user friendly.

In general the developed NESPs were well appreciated by the users or stakeholders, and they recoginised the value of using such environments. However, for its intensive use in the future and to be accepted by the wider audience it needs to create a "culture" of using these applications. Creating such "culture" may happen when stakeholders find the need or urgency to make use of such applications for plan and manage water resources or flood issues. It is also important that such applications should be made known to the wider audience. This can be through social media, city hall website, local newspaper and local weather site.

The workshops with the stakeholders have demonstrated that authorities, water resources mangers and some citizens expressed their interest to extend (additional functionalities or information) or develop similar applications for other study areas. The fact that technical partner (LSBG) in the Alster, Romanian Waters and other stakeholders in Somes Mare and Danube, and the Water Boards in the Noord Brabant expressed particular interest in the developed web and mobile tools and are looking forward for their further development is an indication that such developments may indeed follow.

Although the NESP has been successfully developed and applied to different cases with different participatory process, there were several challenges encountered during its developmental phase. The major challenge in the development of the NEs is the conceptualization of the participatory process and the proper selection and use of technologies.

One of the potential pitfalls in stakeholder participation is the premature termination of the participatory process or its unsatisfactory implementation. The proper design of the different NESPs participatory process was important to maintain participation. The platform structure and information should be clear enough to gain stakeholders' interest. Moreover, as mentioned, for different cases the type of participation can be different given the differences in management strategies, as well as the specific characteristics of case studies.

As presented in this thesis, three types of NESP frameworks were developed and applied for different case studies with different type of participation: (1) information and knowledge sharing (IKS), (2) consultative participation (CP) and (3) collaborative decision making (CDM). The main criterion for selection of a NESP framework is the case study's environmental and management problem and its objectives. Case specific framework based on the type of participation was seen as the most suitable approach in designing the NESP.

The following sub-sections presents conclusions for the different frameworks, judgement engine and NESP information technologies.

8.1.2 NESP-IKS

The framework *NESP-IKS* (**I**nformation and **K**nowledge **S**haring) was conceptualised for effective assimilation of stakeholders' information and knowledge to have a more reliable and up to date information and to improve models and forecasts. The framework has three main components and one optional component: (1) Background information, (2) Information access, (3) Stakeholder participation and (4) Improve model and forecast (optional).

Most of the conclusions for the NESP-IKS application are in general relevant for the other two NESP frameworks applied. This was already presented in the first part of this chapter

It needs to be reiterated here that the Noord Brabant NESP-IKS engaged users (mainly professional users) continuously in the development of the environment. The design and functionality of the environments included user's needs and requirements. Meeting the user's needs and requirements made the platforms more users friendly. This approach in development can be useful for future NESP applications.

In the NESP-IKS different type of data were presented in a way appropriate or understandable for different users (professionals and public). For professional users it was more appropriate to present data in graphs and numerical values and for public users qualitative linguistic form (e.g. 'good', 'bad') was preferred.

8.1.3 NESP-CP

The developed framework termed *NESP-CP* (**C**onsultative **P**articipation) was conceptualised for an effective and efficient consultative participation. The NESP-CP promotes more inclusive stakeholder participation exposed to water related risk. More inclusive participation through consultation can influence existing practices in water resources or flood risk management. Furthermore, it is expected that it will become a valuable system for awareness raising and citizens empowerment in water resources or flood management. The framework has three main components: (1) Risk awareness, (2) Information access and (3) Stakeholder participation.

Same as the NESP-IKS, some of the conclusions for the NESP-CP application are relevant for the other NESP frameworks applied. This was presented in the first part of this chapter. The following paragraphs present a summarised conclusion significant for the development of NESP-CP.

Gaining stakeholders' interest to use the application is a significant challenge. In order to gain stakeholders interest to participate, besides face to face workshops, examples on how to use the system by specific users, such as mayors, city hall employees, environmental agencies should be made available.

Feedback is an important component for the NESP-CP, as demonstrated on the Romanian case studies (Somes Mare and Danube). This feedback can be in different forms, such as text, values, or images. Verification of this feedback should be done before it is published. Such verification can be done manually in the back-end. However with the uncertain number and frequency of feedback it will be difficult to validate this manually. The challenge is to implement a semi-intelligent-automated programme that validates this feedback.

8.1.4 NESP-CDM

The developed *NESP-CDM* (**C**ollaborative **D**ecision **M**aking) framework is intended for participatory collaboration where stakeholders together with experts identify scenarios, realistic management alternatives that address relevant management objectives, rank these alternatives in order of preference by individual stakeholders and aggregate this ranking to represent the view of the whole participating group. When carried out in a fully transparent manner, this can possibly lead to negotiations (amongst stakeholders) towards a consensus on a preferred management alternatives to be implemented. Moreover, the framework considered stakeholder participation in modelling activities (e.g. model validation). Since stakeholders have more knowledge of their local environment, proper assimilation of this knowledge will significantly improve the model results. Moreover, their engagement in modelling can be used as a learning process in understanding more the system and some of the measures introduced. The framework can be summarised as consisting of two main stages: (1) Collaborative modelling and (2) Participatory decision making.

Engaging stakeholders early on the developmental phase is an appropriate approach in developing the NESP-CDM. Incorporating their local knowledge and values early on the stages of developing scenarios and proposed alternatives makes it easier to engage them in the participatory decision making. This has been demonstrated in the Alster and Cranbrook case study. Both case studies, based upon the framework of NESP-CDM, integrated both collaborative modelling and participatory decision making in their participatory process. The case studies results were not conclusive. The number of participating stakeholders was quite small and it did not proceed further with active negotiations. The lack of motivation and shorter participation process may be the reason why actual negotiations stages in collaborative decision making were not reached during this research. The NESP-CDM should be integrated in a longer participatory process and tested with a wider range of stakeholders.

8.1.5 Judgement engine: TOPSIS method

The use of TOPSIS method as the judgement engine for NESP-CDM case studies was useful in formulating judgement on the proposed alternatives, combining the scientific facts with the beliefs and attitudes of the stakeholders in the decision making process. The TOPSIS method provided a close representation of the stakeholders preferences regarding the measures and alternatives. However, for further research it will be of interest to look at other Multi Criteria Decision Methods.

8.1.6   Model uncertainty

One issue that was not fully addressed in the presented case studies is related to limitations of the used models and the associated uncertainty in their results. Elaborate uncertainty analysis that attempts to quantify these uncertainties was not carried out in any of the case studies. Although such analyses can be of value, it is still difficult to communicate its results to non-expert stakeholders. This is certainly a challenge for future similar studies. As an alternative approach to handle and communicate uncertainties, in this study (especially in the Alster case) stakeholders (expert and non-expert) were engaged to validate the model results by comparing them to their knowledge about recent flooding in local areas of the catchment.

8.1.7   NESP information technologies

The selection of *GPL technologies* was critical in building the NESP. Such selection was based on the design of the participatory process and the resources available. Selection of available GPL technologies must be done carefully. In general the use of GPL technologies for such platforms is highly feasible. They do provide the desired level of interactivity in the developed components and have the flexibility to be adopted in other case studies.

Several technologies have been applied and tested in this research. However some of these technologies did not meet the desired functionalities. Developers have to be more resourceful in combining other technologies to meet the desired functionalities. Moreover, it should be also stated that for these kinds of applications programming skills are not sufficient by themselves. When developing platforms for stakeholder participation in water resources or flood risk management interdisciplinary knowledge and skills are needed usually available in teams of developers. For future applications the potential in usage of mobile smartphones may also need to be taken into account for developing applications for stakeholder collaboration

Some visualisations (e.g. swimming pool of alternatives used for the NESP-CDM) have been found less useful by stakeholders, because of their unconventional way of presenting the information. If stakeholders spend more time and become more familiar with this kind of visualisation, their attitude may change.

Mobile technologies, particularly the use of Android platform can be used quite successful for developing water and environmental related applications. The use of mobile applications for water resources and flood management should be encouraged. The use of mobile phones will not only reach the wider audience, users can also instantly provide valuable additional information for decision makers. If these unstructured information (stakeholders observation) were properly assimilated with structured data (measured).

8.2   Recommendations and future work

The following are recommendations to further improve future NESP developements. The presentation is structured as follows: (1) Methods, (2) Web based implementation of water related applications and (3) Group visualisation techniques. Some of the recommendations include a brief literature review to guide researchers interested in this topic:

8.2.1   Methods

Multi criteria decision methods

The TOPSIS method applied in the Alster and Cranbrook case studies provided a close representation of the stakeholders' preferences regarding the measures and alternatives. This method has been widely used and tested in many fields of decision making applications. Worth mentioning is that decision making in the Alster and Cranbrook case studies were based on visual inspection of spatial information presented in the map. Incorporating spatial information in a Multi Criteria Decision Making can be of more value for stakeholders and decision makers. This can be an alternative method to guide and support them on their decisions. Such spatial MCDM is presented below.

The GIS-based Multi Criteria Analysis is an interesting novel method. Developed by Meyer et al. (2007) it was applied as decision support in flood risk management and mapping. The application of MCA (in general) in flood risk management and especially to spatial MCA is still very rare. The resulting study can be categorized as a supporting tool for a long-term planning since it is focused on flood risk management (which involves the risk assessment and evaluation of alternative measures).

The study aimed to improve the flood risk management process in three ways: (1) to consider flood risks which are not measured in monetary terms (2) to show the spatial distribution of these multiple risks (3) to deal with uncertainties in criteria values and to show their influence on the overall assessment.

Uncertainty analysis

In this research elaborate uncertainty analysis that attempts to quantify uncertainties in the models was not carried out in any of the case studies. Although such analyses can be of value, it is still difficult to communicate their results to non-expert stakeholders. This is certainly a challenge for future similar studies. It is recommended to explore (new) possible methods of visualisation techniques to communicate this uncertainty to the non-expert stakeholders.

Game theory

One interesting concept for decision making is Game theory. In definition Game theory is the mathematical study of strategic decision making where individuals or groups (the players) compete and cooperate to resolve conflict. Game theory can in some situations predict people's behaviour, their interest and conflicts during the decision making. Little research has been done in Game theory in application to water resources and flood management. An interesting review by Madani (2009) provides an overview of the applicability of game theory to water resources management and conflict resolution. For further research, it will be interesting to investigate the use of Game theory in the NESP-CDM. Game theory can be an added value for the NESP-CDM negotiation component.

8.2.2   Web-based implementation of water related applications

Semi-distributed Conceptual Models

The presented NESP-CDM collaborative modelling and participatory decision making components made use of pre-prepared model results. Although this approach has been successfully demonstrated, it is interesting to explore possibilities of using web-based (remote) modelling for the collaborative modelling and decision making process. This can be an added value for stakeholder participation.

Flood Forecasting System

Integrating a forecasting system can be a valuable tool not only to gain stakeholders interest in using the application but also for participation. If the models used for forecasting are computationally demanding, Clouds or Grid can be used to simulate the forecasting model(s) and have a cost effective, reliable and efficient forecasting and warning system. Hence, it is recommended to explore effective methods for an integrated forecasting system for the NESP (in particular NESP-IKS).

Data Driven Models

This recommendation is related to flood forecasting. If it is difficult to build a physically based model because of the lack of understanding of the process and its parameters, data driven models (DDM) can be an alternative to predict short term river flows. However, this approach is applicable if there are considerable amount of historical data and the system does not change significantly in time. It is worth stating that a system with series and highly controlled structures can be difficult to be modelled with data-driven. A great deal of research has been done in data driven modelling.Solomatine and Ostfeld (2008) presented past experiences and new approaches in DDM and summarised a number of research challenges in DDM. They have pointed particularly at the use of *hybrid models* (e.g. combining DDM and semi-distributed physically based model; see Corzo et al., 2009) and the inclusion of human expert's knowledge in the modelling process (see Solomatine and Xue, 2004; Corzo et. al., 2007).

If a physical based model is not possible, DDM can be used as an option for NESP-IKS flood forecasting or as a component for the NESP-CDM collaborative modelling. The challenge is to find ways to effectively deploy DDM into a networked environment. Web based DDM can be used by stakeholders (most probably experts) for forecasting and to simulate scenarios of extreme rainfall event (depending on the available historical data of extreme events).

8.2.3 Group Visualisation Techniques

In the participatory decision making of the NESP-CDM, using the so called 'swimming pool of alternatives' stakeholders can keep track on how close are their positions to those of the other participants, which may be supportive for converging to a possible alternative for implementation. Stakeholders have found it less useful compared to other applications because of its unconventional way of presenting the information. This attitude may change if they spend more time and become more familiar with it. This can be a very useful group visualisation tool, however with one limitation: it is almost impossible to visualise individual positions if there are many participants or if there are more than five alternatives. Hence, it is recommended to explore new possible methods to improve this visualisation technique, or perhaps design a new, entirely different kind of group visualisation.

"The overall results of the research demonstrated that networked environments can address the challenges and potential pitfalls in stakeholder participation and enhances participation in water resources and flood management."

ABBREVIATIONS

API	:	Application Programming Interface
AOFD	:	Automatic Overland Flow Delineation
AJAX	:	Asynchronous JavaScript and XML
CSS	:	Cascading Style Sheets
CME	:	Collaborative Modelling Exercise
CP	:	Collaborative Platform
DSS	:	Decision Support System
EC	:	European Commission
FD	:	Flood Directive
FIS	:	Flood Information System
FRM	:	Flood Risk Management
GIS	:	Geographical Information System
GPL	:	General Public License
GPS	:	Global Positioning System
HTML	:	HyperText Markup Language
HTTP	:	HyperText Transfer Protocol
ICT	:	Information and Communication Technologies
KML	:	Keyhole Markup Language
MADM	:	Multi Attribute Decision Method
MCDM	:	Multi Criteria Decision Making
NDDSS	:	Network Distributed Decision Support System
NE	:	Networked Environment
NESP	:	Networked Environment for Stakeholder Participation
NESP-CDM	:	NESP-Collaborative Decision Making
NESP-CP	:	NESP-Consultative Participation
NESP-IKS	:	NESP-Information and Knowledge Sharing
NGO	:	Non Government Organisation
OS	:	Operating System
PHP	:	Hypertext Preprocessor
RDBMS	:	Relational Database Management System
SDI	:	Spatial Data Infrastructure
SQL	:	Structured Query Language
TOPSIS	:	Technique for Order Preference by Similarity to Ideal Solution

URL   : Universal Resource Locator
WaterML   : Water Markup Language
WFD   : Water Framework Directive
WFS   : Web Feature Services
WMS   : Web Mapping Services
WQ   : Water Quality
WRFM   : Water Resources and Flood Management
WRM   : Water Resources Management
XML   : Extensible Markup Language

REFERENCES

Abbott, M.B. (2001). The democratisation of decision-making process in the water sector I. *Journal of Hydroinformatics*, 3(1), pp. 23-34.

Abbott, M.B. and Jonoski, A. (2001). The democratisation of decision-making process in the water sector II. *Journal of Hydroinformatics*, 3(1), pp. 35-47.

Almoradie, A.D.S. (2008). *Hydroinformatics Geo-referenced tools and Web Tachnologies for Flood Management: A Case Study in the Westerschelde, The Netherlands*. Unpublished MSc thesis, UNESCO-IHE Institute for Water Education, Delft.

Almoradie, A., Cortes, J. and Jonoski, A. (2013). Web-based stakeholder collaboration in flood risk management. *Journal of FLOODrisk management*, "in review".

Almoradie, A. (2013). [Assorted images]. Retrieved from https://sites.google.com/site/adrianalmoradie/

Antonopoulou, E., Karetsos, S.T., Maliappis, M. and Sideridis, A.B. (2009). Web and mobile technologies in a prototype DSS for major field crops. *Journal of Computers and Electronics in Agriculture*, doi:10.1016/j.compag.2009.07.024.

Arnstein, A. (1969). A ladder of citizenship participation. *Journal of the American Institute of Planners*, 26, pp. 216-233.

Baker, M., Buyya, R. and Laforenza, D. (2002). Grid and Grid technologies for wide-area distributed computing. *Journal of Software Practice and Experience*, 32(15), pp. 1437-1466.

Bhargava, H.K. and Krishnan, R. (1998). The World Wide Web: Opportunities for Operations Research and Management Science. *INFORMS Journal on Computing*, 10(4), pp. 359-383.

Biggs, S. (1989). *Resource-Poor Farmer Participation in Research: a Synthesis of Experiences From Nine National Agricultural Research Systems*. OFCOR Comparative Study Paper, vol. 3. International Service for National Agricultural Research, The Hague.

Bonn Conference (2001). *Bonn Recommendation for Action. Outcome from the International Conference on Freshwater* [online]. Bonn, Germany. http://www.water-2001.de/outcome/BonnRecommendations/. [Accessed 22 May 2013].

Bousset, J.P., Macombe, C. and Taverne, M. (2005). *Participatory methods, guidelines and good practice guidance to be applied throughout the project to enhance problem definition,colearning, synthesis and dissemination*. Technical Report SEAMLESS Report No.10, Cemagref. SEAMLESS integrated project, EU 6th Framework Programme, contract no. 010036-2.

Centre for Ecology and Hydrology (formerly the Institute of Hydrology), (1999). *Flood Estimation Handbook*. Wallingford, Oxfordshire, UK.

Chase, L.C., Decker, D.J. and Lauber, T.B. (2004). Public participation in wildlife management: What do stakeholders want? *Journal of Society and Natural Resources*, 17, pp. 629-639.

Chess, C., Dietz, T. and Shannon, M. (1998). Who should deliberate when? *Journal of Human Ecology Review*, 5, pp. 60-68.

Choi J., Engel, B. and Farnsworth, L. (2005). Web-based GIS and spatial decision support systems for watershed management. *Journal of Hydroinfromatics*, 7(3), pp. 165-174.

Collectif ComMod, (2006). *Modélisation d'accompagnement. In: Amblard, F., Phan, D. (Eds.),Modélisation et simulation multi-agents*: applications aux sciences de l'homme et de la société. Hermés Sciences, Londres, pp. 217–228.

Corzo, G.A., Solomatine, D.P., De Wit, M., Werner, M., Uhlenbrook, S. and Price R.K. (2009). Combining semi-distributed process-based and data-driven models in flow simulation: a case study of the Meuse river basin. *Journal of Hydrology and Earth System Sciences*, 13, pp. 1619-1634.

Corzo, G.A. and Solomatine, D.P. (2007). Knowledge-based modularization and global optimization of artificial neural network models in hydrological forecasting. *Journal of Neural Networks*, 20 (4), pp. 528-536.

Defra, (2010). Draft strategy for skills and capacity building in local authorities for local flood risk management [online]. UK. http://archive.defra.gov.uk/environment/flooding/documents/manage/surfacewater/capacitybuilding.pdf. [Accessed 22 May 2013].

EC Directive, (2000). *DIRECTIVE 2000/60/EC of the European Parliament and of the Council of 23 October 2000, on the establishing a framework for Community action in the field of water policy* [online]. Official Journal of the European Communities, L 327/1, 22.12.2000, Brussels. http://eur-lex.europa.eu/LexUriServ/LexUriServ.do?uri=OJ:L:2000:327:0001:0072:EN:PDF. [Accessed 22 May 2013].

EC Directive, (2007). *DIRECTIVE 2007/60/EC of the European Parliament and of the Council of 23 October 2007, on the assessment and management of flood risks*[online]. Official Journal of the European Union, L 288/27, 6.11.2007, Brussels. http://eur-lex.europa.eu/LexUriServ/LexUriServ.do?uri=OJ:L:2007:288:0027:0034:EN:PDF. [Accessed 22 May 2013].

Elliott, J., Heesterbeek, S., Lukensmeyer, C.J. and Slocum, N. (2005). *Participatory Methods Tool kit. A practitioner's manual*. King Baudouin Foundation and the Flemish Institute for Science and Technology Assessment (viWTA).

European Union, (2006). *EU Bathing Water Directive* [online]. http://ec.europa.eu/environment/water/-bathing/index_en.html#2006. [Accessed 22 May 2013).

Evers, M. (2008). An analysis of the requirements for DSS on integrated river basin management. *Journal of Management of Environmental Quality*. 19 (1), pp.37 - 53.

Evers, M., Maksimović, Č., Jonoski, A., Lange, L., Ochoa, O., Teklesadik, A., and Makropoulos, C., (2011). *2nd ERA-NET CRUE Research Funding Initiative Flood Resilient Communities – Managing the Consequences of Flooding Final Report* [online]. http://www.crue-eranet.net/Calls/DIANE-CM_frp.pdf. [Accessed 22 June 2013).

Farrington, J. (1998). Organisational roles in farmer participatory research and extension: lessons from the last decade. *Journal of Natural Resource Perspectives*, 27, pp. 1-4.

Fischer, F. (2000). *Citizens experts and the environment. The Politics of Local Knowledge*. Duke University Press, London.

Fischer, F. and Young, J.C. (2007). Understanding mental constructs of biodiversity: Implications for biodiversity management and conservation. *Journal of Biological conservation*, 136, pp. 271-282.

Flood and Water Management Act, (2010). London. Parliament UK.

Foster, I., Kesselman, C., Nick, J.M. and Tuecke, S. (2002). *The Physiology of the Grid: An Open Grid Sevices Architecture for Distributed Systems Integration* [online]. http://www.globus.org/alliance/publications/papers.php [Accessed 22 May, 2013].

Hwang, C.L. and Yoon, K. (1981). *Multiple attribute decision making: methods and applications. Lecture Notes in Economics and Mathematical Systems*, Berlin; New York: Springer-Verlag.

Hickey, S. and Mohan, G. (2005). *Participation: from tyranny to transformation? Exploring new approaches to participation in development*. Zed Books.

ICOL, (2011). *Guidelines for informatics support to awareness raising and reslience enhancement activities*, DIANE-CM, London.

Jonoski, A. (2002). *Hydroinformatics as Sociotechnology: Promoting Individual Stakeholder Participation by Using Network Distributed Decision Support Systems*. Published Phd thesis, UNESCO-IHE Institute for Water Education, Delft.

Jonoski, A., Alfonso, L., Almoradie, A., Popescu, I., van Andel, S.J. and Vojinovic, Z. (2012a). Mobile phone applications in the water domain, *Environmental Engineering and Management Journal*, Volume 11, Issue 5, 2012, Pages 919-930

Jonoski, A., Almoradie, A., Khan, K., Popescu, I. and van Andel, S.J. (2012b). Google Android Mobile Phone Applications for Water Quality Information Management. *Journal of Hydroinformatics*, "in press".

Jonoski, A., Popescu, I., Almoradie, A., Stoica, F., Teodor, S., Jelev, I. and Gorgan, D. (2013). *Deliverable D6.11 Functional prototypes of BSC-OS Flood Portals for citizens: Implementation and Evaluation* [online]. EnviroGRIDS-FP7 European project http://envirogrids.net/index.php?option=com_jdownloads&Itemid=13&view=finish&cid=179&catid=15. [Accessed 22 June 2013).

Khan, K. (2010). *Integrated Web-Mobile Phone Application for Surface Water Quality Monitoring and Alerting Services: Case Study in the Brabantse Delta, Province of Noord-Brabant, The Netherlands.* Unpublished MSc thesis, UNESCO-IHE Institute for Water Education, Delft.

Khelifi, O., Lodolo, A., Vranes, S., Centi, G. and Miertus, S. (2006). A web based decision support tool for groundwater remediation technologies selection. *Journal of Hydroinformatics*, 8(2), pp. 91-100.

Krywkow, J. (2009). *A Methodological Framework for Participatory Processes in Water Resources Management.* Published Phd thesis, University of Twente, Enschede.

Krywkow J., & Hare M., (2008). *Participatory process management.* Proceedings from the iEMSs 2008 International Congress on Environmental Modelling and Software. Barcelona, Catalonia.

Lotov, A.V. (2003). Internet Tools for Supporting of Lay Stakehoders in the Framework of the Democratic Paradigm of Environmantal Decision Making. *Journal of Multi-Criteria Decision Analysis*, 12, pp. 145-162.

LSBG, (2009). *Hochwasserschutz für die Hamburger BinnengewässerBerichte* N3/2009. Hamburg: Landesbetrieb Straßen, Brücken und Gewässer (LSBG).

Madani, K. (2009). Game theory and water resources. *Journal of Hydrology*. 381, pp. 225-238.

Maksimović, Č., Prodanović, D., Boonya-Aroonnet, S., Leitão, J.P., Djordjević, S. and Allitt, R. (2009). Overland flow and pathway analysis for modelling of urban pluvial flooding. *Journal of Hydraulic Research*, 47(4), pp. 512-523.

Malczewski, J. (1999). *GIS and multicriteria decision analysis.* New York.

Mateos, C., Zunino, A. and Campo, M. (2008). A survey on approaches to gridification. *Journal of Software Practice and Experience*, 38(5), pp. 523-556.

Meyer, V., Haase, D. and Scheuer, S. (2007). *GIS-based Multicriteria Analusis as Decision Support in Flood Risk Management* [online]. FLOODsite report T10-07-07 http://www.floodsite.net/html/publications2.asp. [Accessed 22 May, 2013.

Millennial Media (2011). Mobile Mix –The mobile device Index, Q3 2011 [online]. www.milennialmedia.com [Accessed 22 May 2013].

Molkenthin, F., Belleudy, P., Holz, K.P., Jozsa, J., Price, R. and van der Veer, P. (2001). Hydroweb: 'WWW based collaborative engineering in hydroscience'- a

European education experiment in the Internet. *Journal of Hydroinformatics*, 3(4), pp. 239-243.

Munda, G. (1995). Multicriteria *Evaluation in a Fuzzy Environment – Theory and Applications in Ecological Economics*. Heidelberg, Physica Verlag

Paulos, E., Honicky, R. J. and Hooker, B. (2008) *Citizen science: enabling participatory urbanism. In: Handbook of Research on Urban Informatics: The Practice and Promise of the Real Time City*. Information Science Reference, IGI Global.

Pitt, M. (2008). The Pitt Review: Learning lessons from the 2007 floods [online]. Cabinet Office. http://archive.cabinetoffice.gov.uk/pittreview/thepittreview.html [Accessed 22 May 2013]

Pretty, J.N. (1995a). Participatory learning for sustainable agriculture. *Journal of World development*, 23, pp. 1247-1263.

Pretty, J.N. (1995b). *A trainers guide for participatory learning and action*. International Institute for Environment and Development.

Rao, M., Fan, G., Thomas, J., Cherian, G., Chudiwale, V. and Awawdeh, M. (2006). A web-based GIS Decesion Support Systems for managing and planning USDA's Conservation Reserve Program (CRP). *Journal of Environmental Modelling & Software*, 22, pp. 1270-1280.

Rasche, K., Krywkow, J., Newig, J. and Hare, M. (2006). Measuring the intensity of participation along six dimensions. *Proceedings from the Participatory Approaches in Science & Technology (PATH) conference*, Edinburgh.

Reed, M.S. (2008). Stakeholder participation for environmental management: A literature review. *Journal of Biological Conservation*. pp. 2417-2431.

Reed, M.S., Dougill, A.J. and Baker, T. (2008). Participatory indicator development: what can ecologist and local communities learn from each other? *Journal of Ecological applications*, 18, pp. 1253-1269.

Redbridge, (2010). *Redbridge Strategic Flood Risk Assessment* [online]. http://www2.redbridge.gov.uk/cms/planning_land_and_buildings/planning_policy__regeneration/strategic_flood_risk.aspx. [22 May 2013].

Richards, C., Blackstock, K.L. and Carter, C.E. (2004). *Practical Approaches to Participation*. SERG Policy brief No. 1. Macauley Land Use Research Institute, Aberdeen.

Ridder, D., Mostert, A. and Wolters, H. (2005). *Learning together to manage together. Improving participation in water management*. University of Osnabrück, Institute of Environmental Systems Research.

Rowe, G. and Frewer, L. (2004). Evaluating public-participation exercises: A research agenda. *Journal of Science, Technology & Human Values*, 29(4), pp. 512-556.

Simonovic, S.P. (2009). *Managing water resources, Methods and Tools for a Systems Approach*. London: Unesco.

Segura, J.L.A. (2010). *Optimisation of Monitoring Networks for Water Systems: Information Theory, Value of Information and Public Participation*. Published Phd thesis, UNESCO-IHE Institute for Water Education, Delft.

Souchère, V., Millair, L., Echeverria, J., Bousquet, F., Le Page, C. and Etienne, M. (2010). Co-constructing with stakeholders a role-playing game to initiate collective management of erosive runoff risk at the watershed scale. *Journal of Environmental Modelling & Software*, 25, pp. 1359-1370.

Solomatine, D.P. and Ostfeld, A. (2008). Data-driven modelling: some past experiences and new approaches. *Journal of Hydroinformatics*, 10(1), pp. 3-22.

Solomatine, D. P. and Xue, Y. (2004). M5 model trees and neural networks: application to flood forecasting in the upper reach of the Huai River in China. *ASCE Journal of hydrologic engineering*. 9 (6), pp. 491–501

Sultan, N. (2009). Cloud computing for education: A new dawn?. *International Journal of Information Management*, doi:10.1016/j.ijinfomgt.2009.09.004.

Tippet, J., Handley, J.F. and Ravetz, J. (2007). Meeting the challenges of sustainable development- A conceptual appraisal of a new methodology for participatory ecological planning. *Journal of Progress in Planning*, 67, pp. 9-98.

UN-ESCAP, (2003). *Guidelines on participatory planning and management for flood mitigation and preparedness*. Water Resources Series No. 82. United Nations, New York.

Viegas, C., Nunes, S., Fernandes, R. and Neves, R. (2009). Streams contribution on bathing water quality after rainfall events in Costa do Estoril- a tool to implement an alert system for bathing water quality. *Journal of Coastal Research*, SI56, pp. 1691-1695.

Voinov, A. and Bousquet, F. (2010) Modelling with stakeholders. *Journal of Environmental Modelling & Software*, 25(11), pp. 1268-1281.

Wang, L., Simões, N., Ochoa, S., Leitão, J.P., Pina, R., Onof, C., Sá Marques, A., Maksimović, Č., Carvalho, R. and David, L. (2011). An enhanced blend of SVM and Cascade methods for short-term rainfall forecasting. *Proceedings from the12th International Conference on Urban Drainage*, Porto Alegre.

Waterchap Brabantse Delta (2008). *Evaluatie zwemwater Binnenchelde 2008*. Report. In Dutch. English title: Evaluation bathing water Binnenschelde, pp. 1–8.

Waterchap Brabantse Delta (2009). Memo: *Toelichting waterbalansen Markiezaatsmeer en Binnenschelde*. Report. In Dutch. English title: Explanation of the water balance of Lakes Markiezzaats and Binnenschelde, pp. 1–2.

Wates, N. (2000). *The community planning handbook*. Earth scan Publications Ltd, London.

Webler, T., Kastenholz, H. and Renn, O. (1995). Public participation in impact assessment: A social learning perspective. *Journal of Environmental Impact Assessment Review*, 15(5), pp. 443-463.

Whatmore, S.J. and Landström, C. (2011) Flood apprentices: an exercise in making things public. *Journal of Economy and Society*, 40(4), pp. 582-610.

White, I., Kingston, R. and Barker, A. (2010). Participatory geographic information systems and public engagement within flood risk management. *Journal of Flood Risk Management*, 3 (4), pp. 337-346.

WMO, (2006). *Social Aspects and Stakeholder Involvement in Integrated Flood Management*. APFM Technical Document No. 4, Flood Management Policy Series, World Meteorological Organization No. 1008., Geneva

World Bank, (2012). Cities and Flooding: A Guide to Integrated Urban Flood Risk Management for the 21st Century. DOI: 10.1596/978-0-8213-8866-2.

SAMENVATTING

Gebrek aan bewustzijn bij belanghebbenden, betrokkenheid en participatie bij plan- en besluitvormingsprocessen in water- en overstromingsbeheer leidt vaak tot problemen bij de implementatie en acceptatie van de voorgestelde maatregelen. Bewustzijn bij belanghebbenden en participatie zijn cruciaal bij rampenbestrijding en moeten mee worden genomen in elke fase van welk type ramp dan ook. Aangezien de belanghebbenden vaak meer verstand hebben van de echte mogelijkheden en beperkingen van hun omgeving, is bovendien hun betrokkenheid bij planvorming en beheer van cruciaal belang.

Belanghebbenden kunnen worden ingedeeld in categorieën zoals overheidsinstanties, gemeenschappen in gebieden met een overstromingsrisico, NGOs, riviercommissies, private sector, en wetenschappelijke groepen. Het delen van informatie en herhaaldelijke interactie tussen belanghebbenden zijn nodig om vertrouwen op te bouwen, te onderhandelen over de best mogelijke voorwaarden, en om samenwerking tussen jurisdicties en sectoren te bevorderen. De centrale uitdaging bij participatie door belanghebbenden is het opstarten en het onderhouden van het participatieve proces. Ruimtelijke spreiding en een verscheidenheid aan (zelfs tegengestelde) belangen van belanghebbenden kunnen naar voren komen als obstakels in het onderhouden van het participatieve proces.

In dit onderzoek, getiteld: "Netwerkomgevingen voor participatie van belanghebbenden (NESP) in water- en overstromingsbeheer", wordt een aantal van deze uitdagingen en obstakels geadresseerd. Netwerkomgevingen (NE) zijn op internet gebaseerde computerondersteunde of mobiele omgevingen waarmee deelnemende entiteiten, zoals belanghebbenden, op afstand virtuele interactie kunnen hebben. NESP is bedoeld om belanghebbenden te laten participeren in water- en overstromingsbeheer door uitwisseling van informatie, planvorming, onderhandelen en besluitvorming te ondersteunen.

De recente ICT (informatie en communicatie technologie) ontwikkelingen bieden innovatieve oplossingen voor het ontwikkelen van de NESP. Vanaf het begin van het computer- en internettijdperk is het World Wide Web in onze maatschappij in toenemende mate gebruikt als een technologie om toegang te krijgen tot informatie en voor communicatie tussen organisaties en individuen. Mobiele technologie heeft bovendien aangetoond nog meer voordelen te hebben bij het bereiken en betrekken van het grootste deel van de bevolking en potentiële belanghebbenden. De vooruitgang van mobiele technologie en de ontwikkeling van mobiele applicaties maken de weg vrij om mobiele technologie te gebruiken voor het verzamelen van gegevens, voor het op afstand uitvoeren van modellen, en voor het verspreiden van informatie. Feitelijk vormen het Internet, World Wide Web, en mobiele en draadloze technologieën een krachtige omgeving om de in deze studie voorziene NESP te ontwikkelen en toe te passen.

De belangrijkste doelstelling van dit werk is onderzoek te doen naar conceptualisering, ontwerp, en implementatie van innovatieve op internet gebaseerde en mobiele omgevingen voor participatie van belanghebbenden, en daarbij gebruik te maken van recente ontwikkelingen in de ICT. Het onderzoek integreert nieuwe methodes voor participatie van belanghebbenden in alle fasen van projectplanning en analyse, inclusief ondersteuning bij onderhandelingen om tot 'win-win'-oplossingen te komen.

Om de drie vormen van participeren: (1) delen van informatie en kennis (IKS), (2) participatie op basis van raadplegen (CP), en (3) besluitvorming door samenwerken (CDM), te behandelen, zijn drie situatieafhankelijke NESP raamwerken geconceptualiseerd. Verwijzend naar de drie vormen van participatie, zijn de raamwerken als volgt genoemd en afgekort: (1) NESP-IKS, (2) NESP-CP en (3) NESP-CDM.

Het NESP-IKS raamwerk (delen van informatie en kennis) was geconceptualiseerd om kennis en informatie van belanghebbenden effectief te kunnen verwerken in water- en overstromingsbeheer. Dit kan resulteren in de mobilisatie en het benutten van betrouwbaardere en actuelere kennis bij processen in het water- en overstromingsbeheer. Daarbij biedt het raamwerk vakmensen de mogelijkheid om waarnemingen van belanghebbenden te gebruiken om hun modellen en voorspellingen te verbeteren. Het conceptuele raamwerk bestaat uit drie hoofdonderdelen en een optioneel onderdeel: (1) achtergrondinformatie, (2) toegang tot informatie, (3) participatie van belanghebbenden, en (4) verbetering van modellen en voorspellingen (optioneel).

Het NESP-CP raamwerk (participatie op basis van raadplegen) was geconceptualiseerd voor een effectievere en meer inclusieve manier van participeren. Een meer inclusieve participatie door raadplegen kan bestaande werkwijzen in water- en overstromingsbeheer beïnvloeden. Van de ontwikkelde NESP-CP toepassing wordt verwacht dat het een waardevol systeem is om bewustzijn en slagkracht van belanghebbenden in water- en overstromingsbeheer te vergroten. Het raamwerk heeft drie belangrijke componenten: (1) Risicobewustzijn, (2) toegang tot informatie, en (3) participatie van belanghebbende.

Tot slot, is het NESP-CDM raamwerk (besluitvorming door samenwerken) bedoeld voor een samenwerkende vorm van participatie waarbij belanghebbenden samen met vakmensen relevante scenario's en realistische beheeralternatieven identificeren die aan samen overeengekomen doelstellingen tegemoet komen. Vervolgens beoordelen de participerende partijen de geprefereerde alternatieven: eerst door individuele belanghebbenden in staat te stellen hun eigen rangschikking te maken van de alternatieven, hetgeen dan vervolgd wordt door de rangordes samen te voegen om zo de voorkeur van de participerende groep als geheel te laten zien. Indien dit proces wordt toegepast op een volledig transparante manier, dan kan het mogelijk leiden tot onderhandelingen (tussen de belanghebbenden) om tot een gezamenlijke voorkeur te komen voor welk beheeralternatief wordt toegepast. Het raamwerk voorziet ook in de participatie van de betrokkenen bij het modelleringproces (bijvoorbeeld voor modelvalidatie). Aangezien belanghebbenden meer kennis hebben van hun lokale omgeving, kan het juiste gebruik van deze kennis modelresultaten aanzienlijk verbeteren. Daarbij kan hun deelname aan het modelleringproces, dit wordt 'Collaborative modelling' (samenwerkend modelleren) genoemd, worden gebruikt als een leerproces voor een beter begrip van het systeem en van de daarin voorgestelde maatregelen. Het raamwerk kan worden samengevat middels de twee fasen waaruit het bestaat: (1) Collaborative modelling en (2) participerende besluitvorming.

De voornaamste criteria bij het kiezen van een NESP raamwerk zijn projectafhankelijk op basis van omgevingsfactoren, het type beheersvraagstuk en de bijbehorende doelstellingen. Daarbij is het belangrijk om bij het kiezen van het NESP raamwerk eerst de karakteristieken van de praktijkstudie te bepalen, alvorens het participatieproces te ontwerpen. Deze bepaling zal ook richting geven aan het opzetten en implementeren van de NESP.

De drie NESP raamwerken zijn gebruikt om applicaties te ontwikkelen en te testen voor 5 praktijkstudies met verschillende milieuproblemen en beheerdoelstellingen. De praktijkstudies zijn (1) Plassen en meren in Noord-Brabant, Nederland, (2), het stroomgebied van de Somes Mare, Roemenie, (3) het stroomgebied van de Donau (het Braila-Isaccea gebied), (4) het stroomgebied van de Cranbrook, Londen, Verenigd Koninkrijk en (5) het stroomgebied van de Alster, Hamburg, Duitsland.

Het NESP-IKS raamwerk is toegepast in de praktijkstudie van Noord-Brabant. De doelstelling daar is om verschillende gebruikers, zoals badgasten en surfers, te voorzien van actuele informatie over waterkwaliteit van plassen en meren. Het NESP-CP raamwerk is toegepast in zowel de Donau als de Somes Mare praktijkstudie. Beide praktijkstudies gaan over overstromingsproblematiek, en de doelstelling is om verbeterd overstromingsbeheer te bewerkstelligen door het bewustzijn te vergroten bij organisaties die verantwoordelijk zijn voor het waterbeheer, vakmensen, en bredere belangengroepen en burgers, en met deze groepen onderling informatie te delen en te verspreiden. NESP-CDM is toegepast in de praktijkstudies van de Cranbrook en Alster waar de vergelijkbare doelstelling gold om belanghebbenden een stem te geven in de plan- en besluitvorming voor het omgaan met overstromingsrisico's.

Voor de praktijkstudie van Noord-Brabant is een geïntegreerde internet en mobiele telefoon applicatie ontwikkeld en toegepast, terwijl in de andere vier praktijkstudies internet applicaties zijn toegepast. In alle praktijkstudies zijn deze applicaties ontwikkeld en getest middels bijeenkomsten met eindgebruikers en belanghebbenden. Meestal werd de ingebruikname van de NESP gedaan door middel van een bijeenkomst, om zo aan de belanghebbenden de NESP toepassing te kunnen introduceren en demonstreren. Het aantal vervolgbijeenkomsten hing vervolgens af van het type en de mate van participeren. Na afloop kregen belanghebbenden de tijd om de applicaties te gebruiken en te testen. Tot slot zijn ze tijdens de afsluitende bijeenkomsten gevraagd hun evaluatie van de applicaties te rapporteren.

Over het algemeen werden de ontwikkelde NESPs positief ontvangen en de gebruikers / belanghebbenden zagen duidelijk meerwaarde in het gebruik van dergelijke netwerkomgevingen. Waterbeheerders, beleidsmakers, en enkele belanghebbenden en burgers kwamen met het verzoek om de NESP toepassingen te verlengen, extra informatie toe te voegen, of om soortgelijke applicaties te ontwikkelen voor andere praktijkstudies in hun gebied.

In dit onderzoek zijn voor de ontwikkeling van de applicaties verschillende technologieën getest en toegepast. General Public License (GPL) technologieën zijn uitgebreid gebruikt voor de ontwikkeling van applicaties. De keuze van de specifieke GPL technologie is cruciaal bij het ontwikkelen van de NESP. De selectie hing af van het ontwerp van het participatieproces en van de beschikbare middelen. In concrete termen heeft dit onderzoek aangetoond dat de selectie van GPL technologieën zorgvuldig moet worden gedaan op basis van de volgende criteria: (1) de toepasbaarheid in het raamwerk, (2) flexibiliteit en compatibiliteit met andere technologieën, (3) voor de reeds bestaande componenten, moet gekozen worden voor applicaties waarmee de belanghebbenden bekend zijn (bijvoorbeeld Google maps), (4) het gebruiksgemak van de technologie en (5) de technologie moet breed worden ondersteund door de internationale gemeenschap en continu verder worden ontwikkeld.

Over het algemeen is het gebruik van GPL technologieën voor dergelijke netwerkomgevingen heel goed toepasbaar. Ze bieden het gewenste niveau van interactiviteit in de ontwikkelde componenten en ze hebben de flexibiliteit om ook in andere praktijkstudies te worden toegepast. Het moet echter worden opgemerkt dat voor dit soort applicaties, programmeervaardigheden op zichzelf niet voldoende zijn. Voor het ontwikkelen van netwerkomgevingen voor participatie van belanghebbenden in waterbeheer en omgaan met overstromingsrisico's moeten interdisciplinaire kennis en vaardigheden aanwezig zijn. Deze vindt men doorgaans alleen in teams van ontwikkelaars met verschillende achtergronden.

Het in dit onderzoek gepresenteerde werk heeft aangetoond dat NESP, zoals op internet gebaseerde en mobiele applicaties, de potentie hebben om de obstakels bij participatie van belanghebbenden in water- en overstromingsbeheer te overwinnen.

Adrian Delos Santos Almoradie
Delft, Nederland

ACKNOWLEDGEMENT

This research study could have not been possible without the support of UNESCO-IHE, Hydroinformatics chair group, PhD colleagues and friends. I really appreciate the trust, patience, kindness, motivation, time and the scientific and personal discussions we had during the 4 years I spent as a PhD researcher.

My sincerest gratitude goes to my supervisor Dr. Andreja Jonoski for giving me an opportunity to work on this research. This research helped me expand my horizon as a hydroinformatician. His vast knowledge and wisdom, guidance, patience, kindness and trust became a beacon of light during my struggles as a PhD researcher. He shared with me bright ideas and understanding on sociotechnology in the field of hydroinformatics. His principles somehow also influenced me to become a better person, as well as a better scientist.

I am also truly grateful to my promotor Prof. Dimitri Solomatine for his guidance during my PhD research. I still remember the time when he gave me an opportunity to do my MSc research on web-based technologies for flood management; it has been a stepping stone for my PhD research. During my PhD research he has become a source of inspiration, always reminding me to do something new for science.

I am also extremely grateful to Dr. Ioana Popescu. Her support, guidance and encouragement gave me strength during the stressful time of my PhD.

I would like to thank Dr. Schalk Jan van Andel for giving me an opportunity to work on one of the project case studies of this research. His guidance has been valuable in this research. Also my gratitude to him for translating parts of this dissertation. To Dr. Biswa Bhattacharya, Dr. Giuliano Di Baldassarre, Jos Bult- thank you for proof reading the translated version, Jolanda Boots and the hydronformatics staff - thank you for those easy moments and bright ideas you have shared.

My special thanks to Dr. Gerald Corzo for the guidance on information technology. Those basketball days, playing together with our friends are worth to remember, as well as the time UNESCO-IHE basketball team won the championship against all odds. To former MSc. students I co-supervised, Juliette Cortes, Kamruzzaman Khan and Bin Chai - my thanks for the contribution to this research.

My deepest gratitude also goes to the EnviroGRIDS, LENVIS and DIANE-CM projects which have funded this research, especially to the project partners that have contributed to this research with the work on these projects.

I am thankful to my UNESCO-IHE friends, Michael Siek, Micah Mukolwe, Anuar md Ali, Mijail Arias Hidalgo, Blagoj Delipetrev, Girma Yimer, Tracy Duong, Mario Castro and Nagendra Kayastha for sharing experiences and advices as PhD researchers and for those pleasant and cheerful conversations.

To Delft Association: Chris Ornum, Jesus Quino, Loreen Villacorte, Rose Marie Salazar, Nuttakan (Aim) Wongfun, Raquel Dos Santos, Navchaa Tugjamba and Anoja Kaluarachchi and to Delft TGIF: Arturo Fernando, Anabel Anarna, Aymeeh Diaz, Luchie Dela Torre and Marceliano Carlota - thank you for the great time and friendship.

To Ams and Adrian Huising and Lucille and Arjen Pilot - many thanks for the friendship and thoughts, my family had a great stay in the Netherlands. My special thanks to Ate Maria Vink and family and Mona Delos Reyes, you have been a family to me in the Netherlands.

My deepest thanks to University of San Carlos Water Resources Center, Fr. Herman Van Engelen, Fe Walag and Nenita Jumao-as for encouraging me to take a masters programme at UNESCO-IHE, without which I would not be here where I am now.

I am truly grateful to my parents for having education as the top priority of the family. To my parents in-law Merinisa and Rodigilio - thank you for your support and for looking after my wife and son in the Philippines during my masters and the first 2 years of my PhD. To my wife - thank you for the encouragement, understanding and patience. To my kids Ian and Iliana - you have been the greatest source of inspiration and strength to complete this book.

Lastly, I would like to thank all members of the doctoral committee for their review and the provided comments. The comments and review was really valuable for the quality and completion of this book.

ABOUT THE AUTHOR

Adrian Delos Santos Almoradie was born on the 18th of October 1981 in Masbate, Masbate, Philippines. After finishing Bachelors of Science in Civil Engineering in mid-year of 2003 in the University of San Carlos - Technological Center (Cebu City), he took the Philippine professional licensure civil engineering exams with outstanding marks.

As a Civil Engineer he expected to work in the structural and construction industry or consultancy. However in 2004, he got his first job as a Hydrological engineer in the Water Resources Center (WRC) of University of San Carlos. Working for 2 years in WRC, he gained more knowledge about the water cluster of Civil Engineering. In year 2006, he enrolled in the Masters of Science programme at UNESCO-IHE, Delft the Netherlands. After two years, in 2008 he obtained an MSc degree in Water Science and Engineering -Hydroinformatics specialization.

He worked for a year in a special program of the Hydroinformatics chair group at UNESCO-IHE, before starting his PhD in 2009. His PhD task included co-supervision of MSc students during their thesis and some lecturing or seminars. During his MSc and PhD studies, he expanded his research interest on flood risk and water resources management, stakeholder participation, multi-criteria analysis, decision support systems, hydrology-hydrometry, hydrological (surface and groundwater) and flood modelling, uncertainty analysis, GIS, web-based GIS, Spatial and Temporal Data Infrastructure (STDI), database management system and development of web-based computer and mobile applications for water resources management.

Currently he is working as a Post-Doc with the Water Profile- Water resources management and Eco-Hydrology group of Prof. Mariele Evers in the Department of Geography, University of Bonn, Germany.

Scientific publications

Peer Reviewed International Journals

Almoradie, A., Cortes, J. and Jonoski, A. (2013). Web-based stakeholder collaboration in flood risk management. Journal of Flood Risk Management, DOI: 10.1111/jfr3.12076

Almoradie, A., Jonoski, A., Stoica, F., Solomatine, D. and Popescu, I. (2013). Web-based flood information system: case study of Somes Mare, Romania. Environmental Engineering and Management Journal, Volume 12, Issue 5,Pages 1065-1070.

Almoradie, A., Popescu, I., Jonoski, A. and Solomatine, D. (2013), Web Based Access to Water Related Data Using OGC WaterML 2.0. International Journal of Advanced Computer Science and Applications(IJACSA), EnviroGRIDS Special Issue on "Building a Regional Observation System in the Black Sea Catchment", http://dx.doi.org/10.14569/SpecialIssue.2013.030310

Evers, M., Jonoski, A., Maksimović, Č., Lange, L., Ochoa, O., Teklesadik, A., Cortés, J., Almoradie, A., Simões, N. E., Wang, L., and Makropoulos, C., (2011). Collaborative modelling for active involvement of stakeholders in urban flood risk management. Natural Hazards and Earth System Sciences, 12, 2821-2842

Jonoski, A., Alfonso, L., Almoradie, A., Popescu, I., van Andel, S.J., and Vojinovic, Z. (2012). Google Android Mobile Phone Applications for Water Quality Information Management. Journal of Hydroinformatics, 11(5), 919-930

Jonoski, A., Almoradie, A., Khan, K., Popescu, I. and van Andel, S.J. (2012). Google Android Mobile Phone Applications for Water Quality Information Management. Journal of Hydroinformatics, 15(4), pp. 1137-1149.

Conference Proceedings

Almoradie, A., Jonoski, A., Stoica, F., Solomatine, D. and Popescu, I. (2012). *Web-based Flood Information System: Case study of Somes Mare, Romania*. Proceedings from International Conference „ECOIMPULS 2012 - Environmental Research and Technology". Regional Business Center Timisoara, Romania.

Popescu, I., Almoradie, A. and Jonoski, A. (2012). *Environmental research in the Black Sea Catchment: Flood modelling case studies in Romania*. Proceedings from BALWOIS conference 5th Conference on Water Climate and Environment. Ohrid, Republic of Macedonia.

Jonoski, A., and Almoradie, A. (2012). *Web-based tools for collaborative flood risk management*. Proceedings from the 10th International Hydroinformatics Conference, Hamburg, Germany.

Ochoa, S., Evers, M., Jonoski, A., Maksimovic, C., Almoradie, A., Cortes, J., Makropoulos, C., Dinkneh, A., Simoes, N., Wang, L., van Andel, S.J. and Osmani, S. (2011). *Enhancement of urban pluvial flood risk management and resilience through collaborative modelling: a UK case study*. Proceedings from the 12th International Conference on Urban Drainage. Porto Alegre, Brazil.

Cortes, J., Almoradie, A., Jonoski, A., van Andel, S.J., Evers, M., Langue, L., Dinkneh, A., Maksimovic, C., Ochoa, S., Simoes, N., Wang, L., Osmani, S. and Makropoulos, C. (2011). *Flood risk management via collaborative modelling. Proceedings from the Computing and Control for the Water Industry Conference*. University of Exeter, United Kingdom.

Khan, K., Almoradie, A. and Jonoski, A. (2011). *Integrated Web-Mobile Phone Application for Water Quality Monitoring and Alerting Services*. Proceedings from the 3rd International Conference on Water and Flood Management. Dhaka, Bangladesh.

Almoradie, A., Jonoski, A., Xuan, Y., Gichamo, T. and Solomatine, D. (2010). *Web-based solutions for flood risk analysis, modelling and management*. Proceedings from the 9th International Hydroinformatics Conference, Tianjin, China.

Jonoski, A. and Almoradie, A. (2010). *Google Android mobile phone demonstration applications for water quality information dissemination*. Proceedings from the 9th International Hydroinformatics Conference, Tianjin, China.

Jonoski, A., van Andel, S.J., Popescu, I. and Almoradie, A. (2010). *Distributed Information Systems Providing Localised Environmental Services for All: Case Study on Bathing Water Quality in The Netherlands*. Proceedings from the BALWOIS Conference, Ohrid, Republic of Macedonia.

Jonoski, A., Popescu, I., Almoradie, A. and Mens, M. (2009). *Web-based flood management knowledge dissemination using Google Maps and Google Earth platforms*. Proceedings from the 7th ISE and 8th International Hydroinformatics Conference, Chile.